The history of Bob Cook's successful experiments with inertial propulsion systems.

DEDICATION

This book is dedicated to all those who contributed to the Cooks' significant research, and to those who have made this book a reality.

John W. Campbell Jr. articles reprinted by permission of the author and the author's agents, Scott Meredith Literary Agency, Inc., 845 Third Avenue, New York, New York 10022

Originally published in the United States by CIP Systems, Inc., 1980.

Updated, edited, and republished by CID Engine Press, 2012.

10 9 8 7 6 5 4 3 2 1
Second Edition

Library of Congress Number 80-067979.

ISBN: 0-9604584-0-9.

Cover by: Rob Cook, Jr.
Design by: Gordon Chun and Rob Cook, Jr.
Cartoons and Illustrations by: Chas Fleischman
Photography by: Robin Rothstein and Rob Cook, Jr.
Please direct all mail inquiries to CID Engine Press, 209 Pineview Drive, Pickens, SC 29671.

For more information visit: http://cforce.cc

The CID Engine Project

The DEATH of ROCKETRY

by Joel Dickinson with Bob Cook

TABLE of CONTENTS

1

INTRODUCTION

There now exists an invention capable of profoundly influencing the future of technology. This is a new and unique propulsion system that can convert energy to a working force far more efficiently than anything in use today, and which has already been tested and validated.

This book evidences that a reactionless drive system has in fact been demonstrated to work successfully, thereby signaling the death of rocketry. The man who first accomplished this is Bob Cook, an inventor who holds two United States patents for his reactionless drive principles. Cook's achievement involves the synthesis of ideas generated from several acts of insight as well as from many years of experiments.

Our present world faces an overwhelming energy crisis. This book, the story of Bob Cook and his breakthrough, the Cook Inertial Propulsion (CIP) engine, provides a solution. The answers *are* here.

Joel Dickinson, author, with the inventor, Bob Cook.

2

THE REACTIONLESS DRIVE

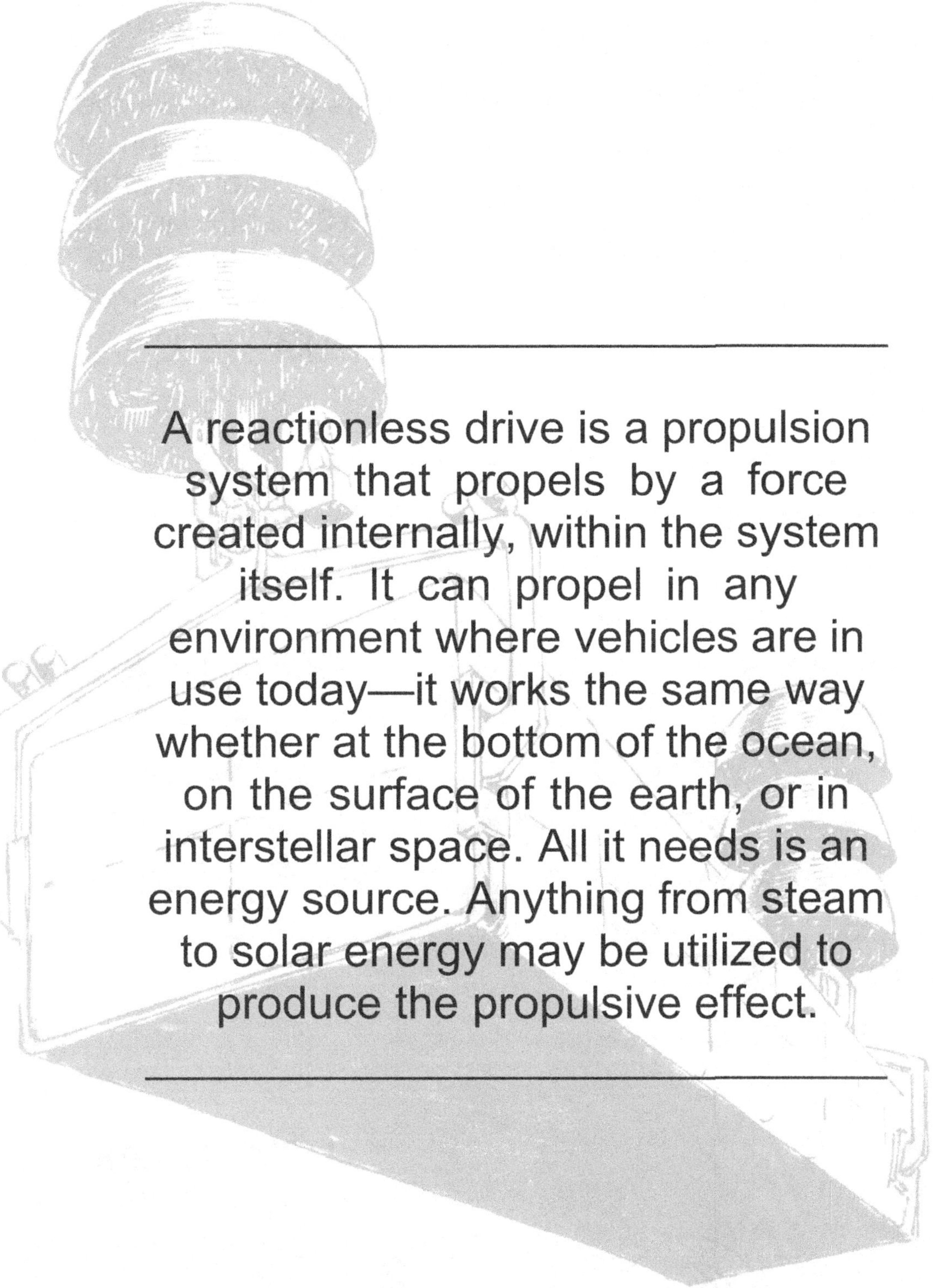

A reactionless drive is a propulsion system that propels by a force created internally, within the system itself. It can propel in any environment where vehicles are in use today—it works the same way whether at the bottom of the ocean, on the surface of the earth, or in interstellar space. All it needs is an energy source. Anything from steam to solar energy may be utilized to produce the propulsive effect.

Imagine a car using this kind of drive. There would be no need for a transmission or drive train to turn the wheels (Fig. 2-1a). Instead, the engine would produce a thrust *within* the car. In this system, all the wheels are needed for are to give the car something to roll on (Fig. 2-1b). You could operate such an auto anywhere on any surface.

By directing your force upward you could lift the vehicle as high off the ground as you wished, then let it down fast or gently. If you accelerated a reactionless drive vehicle at 1 *g*, it would go 0-65 MPH in 3 seconds flat—or, if you reversed the force, braking in the same 3 seconds, even on the slickest patch of ice.

Fig. 2-1

(a) Reaction drive;
(b) Reactionless drive.

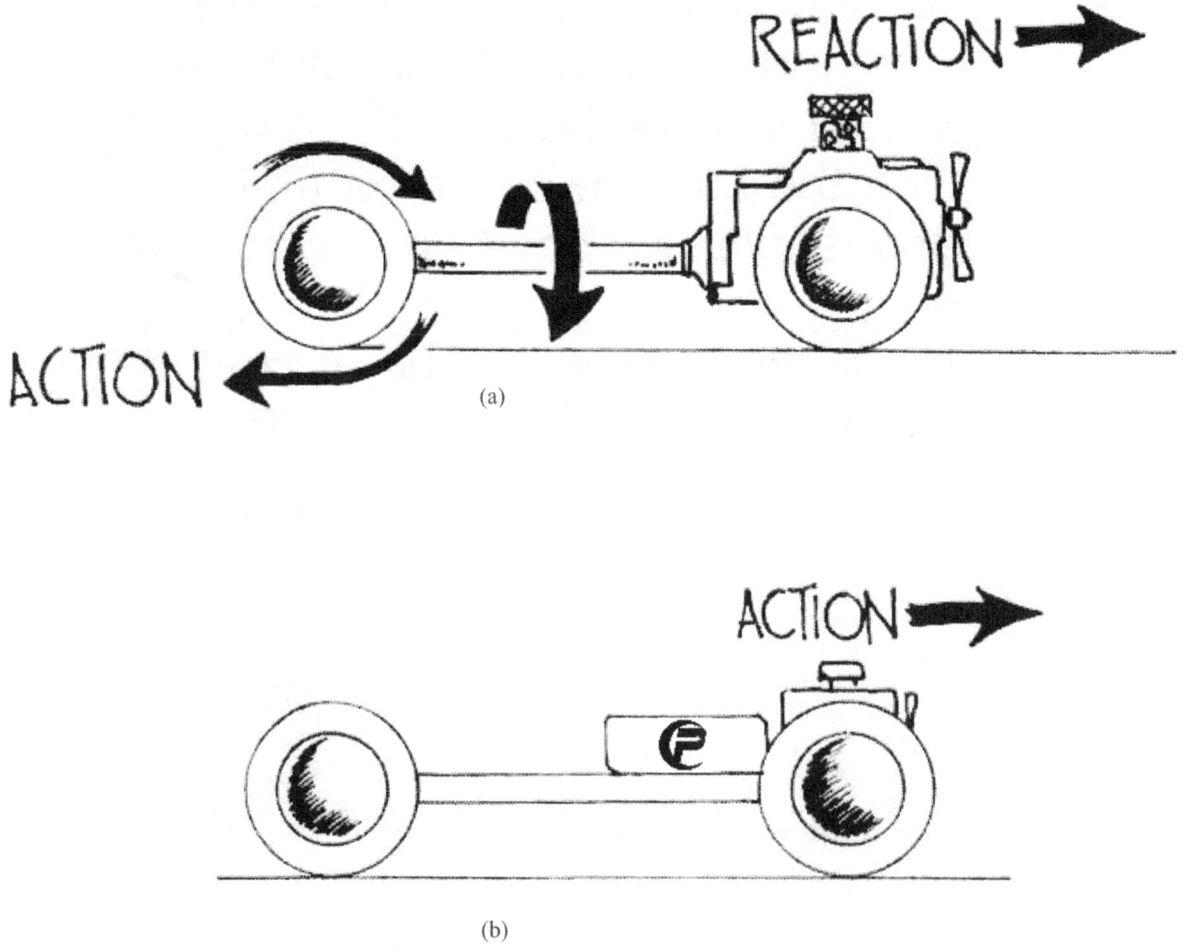

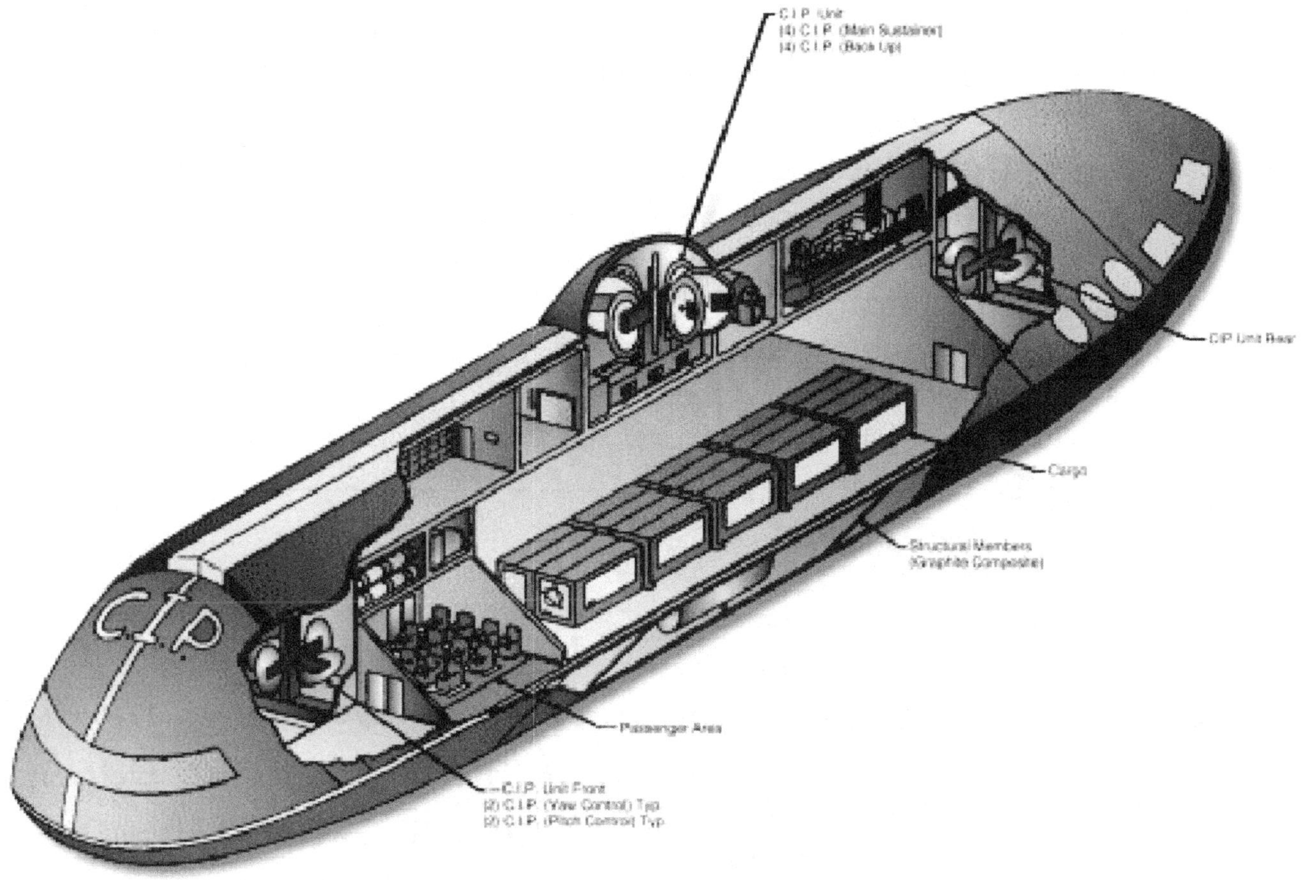

Fig. 2-2

An aerospace transport powered by reactionless drive.

Now imagine an aircraft powered by this.

In a conventional system, the propellant that lifts the plane is air. Wings require a large surface area to generate lift due to the low density of air, which gets lower the higher you go until this lift function is lost.

With the reactionless drive, however, lift is produced within, so wings and propellers are completely unnecessary. This allows for a narrower silhouette with less drag (Fig. 2-2). At higher altitudes friction and gravitational resistance lessens, resulting in greater efficiency. True hovering craft would be perfectly feasible. Such ***forceborne*** aircraft could stop in midair and yet hold altitude.

Let us compare the capabilities of several current space propulsion systems with the reactionless drive.

Presently, the chemical rocket is the prime means of launching from earth, but its extremely low efficiency makes it a less than satisfactory method of travel. (For example, the space shuttle rocket engines were about 16% energy efficient for launching the orbiter into space.) This inefficiency prevents it from accelerating for more than a short period.

Two classes of nuclear rockets have been considered. One, the fission type, utilizes a uranium-fueled, solid-core reactor that heats a working fluid, such as hydrogen, that then accelerates through a nozzle as in the chemical rocket. The other type of nuclear rocket uses energy from the decay of radioactive materials, an engine primarily suitable for low thrust. Safety provisions necessary to protect crew, service personnel, and equipment from excessive hard radiation have caused considerable and expensive delays in such projects.

Three kinds of electrical propulsion have also been utilized. In electrothermal propulsion a working fluid (such as nitrogen or hydrogen) is passed over hot metal surfaces and then expanded in a supersonic nozzle. Electromagnetic propulsion for flight vehicles harnesses motive power produced by high speed discharge of plasma. Along with electrostatic (also known as ion) propulsion, electromagnetic propulsion is capable of attaining specific impulses exceeding those of thermal propulsion devices by a considerable margin.

Presently, the chemical rocket is the prime means of launching from earth, but its extremely low efficiency prevents it from accelerating for more than a short period.

Still another technique involves using the radiation pressure of photons. With this method, a solar sail functions as a reflector upon which the bouncing of photons creates a reaction force that generates propulsion. A clipper sailing through the heavens!

Fig. 2-3

Artificial gravity produced by acceleration and deceleration at 1 *g*.

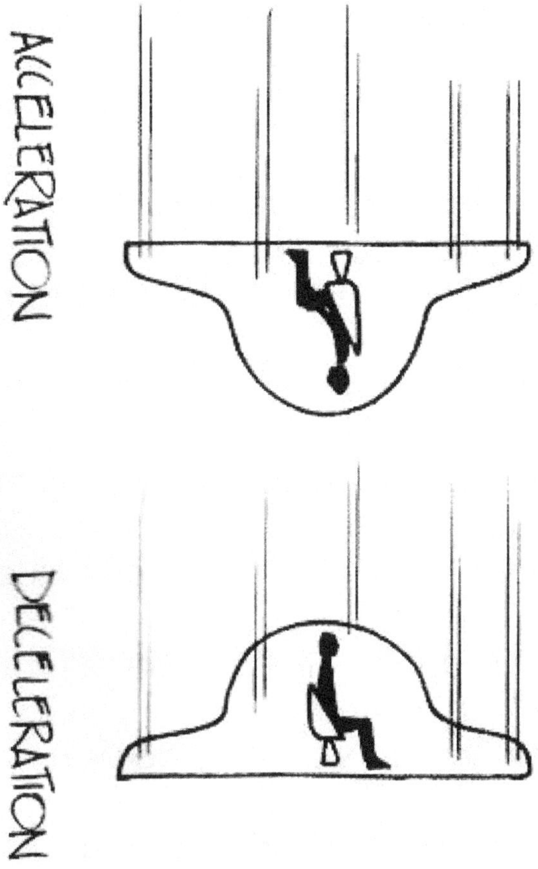

In contrast, a reactionless drive will do things in space beyond the ability of any propulsion system available today.

Since the energy needs of a reactionless drive are so exceedingly small, it could continuously accelerate indefinitely. After 24 hours of acceleration at 1 *g* a spacecraft would be moving at almost 2 million MPH and still gaining in speed. At this rate a trip from Earth to Mars would take less than three days (when Mars is closest) or about five days (when Mars is at the far side of the sun). This rate of acceleration also duplicates the effects of normal gravity, so passengers could not distinguish between sitting in the ship or relaxing at home.

Chemical rockets cannot even approach this potential. There is no need for radiation shielding as with nuclear propulsion. And when compared with electrical propulsion, the force potential of the reactionless drive is a brute—it's like using a smoke machine to push back against a bulldozer (Fig. 2-4).

What influence would the reactionless drive have on our planet and our lives?

All transportation costs would drop. Flying would not only be cheaper but faster, smoother, and super safe. Commuting, no longer limited by narrow roads and bridges, would be far easier.

Shipping goods to anywhere in the world would be a swift matter, especially useful in times of emergency. Delivery vehicles could unload their cargo anywhere, whether the 36th floor of a skyscraper or the top of a distant mountain (Fig. 2-11). Ships with units in the stern and bow could actually direct the force so the ship could dock sideways (Fig. 2-5).

Strap a lightweight "force pack" on your back and you can float with the breeze, glide with the birds, and drift with the clouds (Fig. 2-6). Traveling adventurers and photographers could easily reach isolated locations. Fishing enthusiasts could drop in on some virgin lake to make that last lucky catch of the day.

Portable floodlights could be flown in to light construction projects, sporting events, and disaster areas. Camera crews could get fantastic vantage points riding on or remotely controlling special units.

Fig. 2-4

Ion propulsion versus the reactionless drive is like a smoke machine versus a bulldozer.

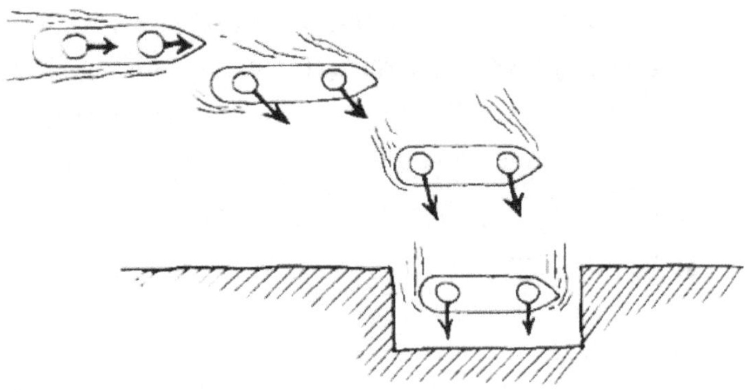

Fig. 2-5

Ship berthing without the aid of tugboats.

Fig. 2-6

Commuting by reactionless drive.

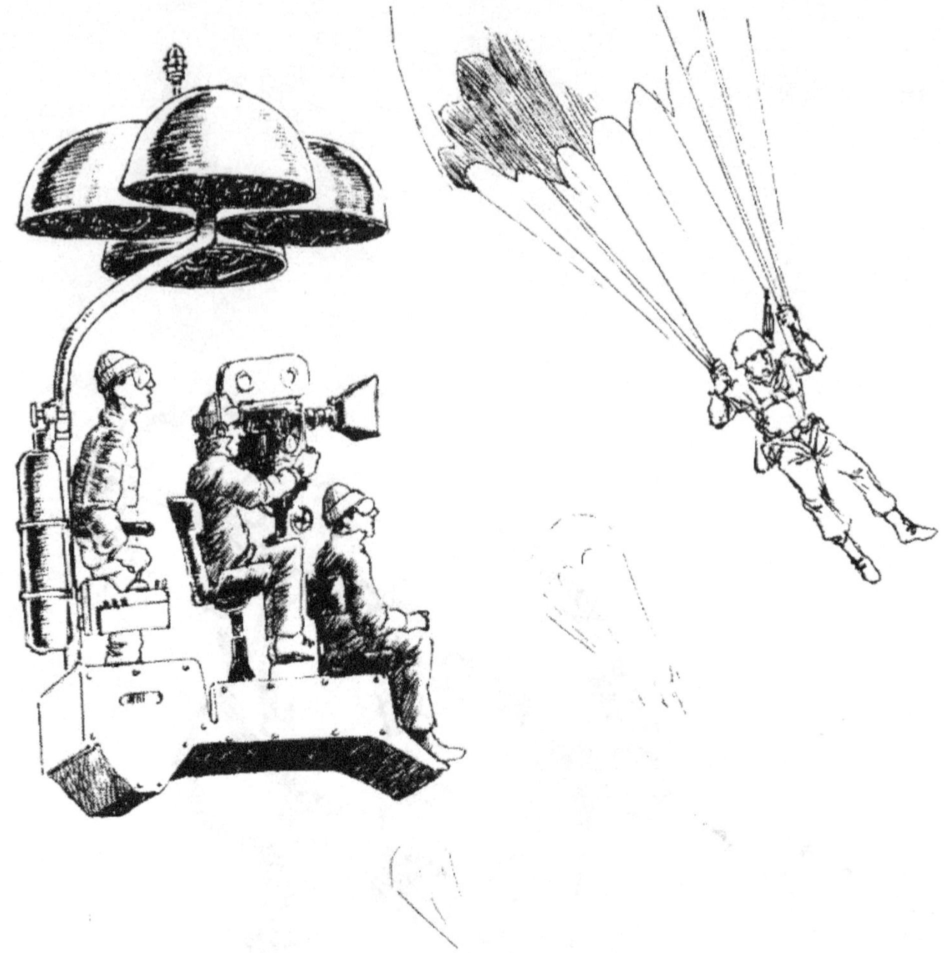

Fig. 2-7

Filmmaking will be revolutionized.

Platforms mounted with these units could hover close to buildings for painting, changing light bulbs, or washing windows. Fire, police, and medical personnel could more effectively provide emergency services with their new vehicles (Fig 2-8).

Powerful units could be used to relocate entire buildings. Houses could be constructed with specialized foundations that could allow them to be lifted and flown to just about anywhere (Fig. 2-9). Instead of packing up and renting a vacation home, just take your own residence and personal belongings with you. Large office buildings and skyscrapers could be built outside a city and then moved into place, eliminating urban construction site clogging and noise.

Fig. 2-8

(a) Firefighters will gain new capabilities.
(b) Sea rescue facilitated by reactionless drive.

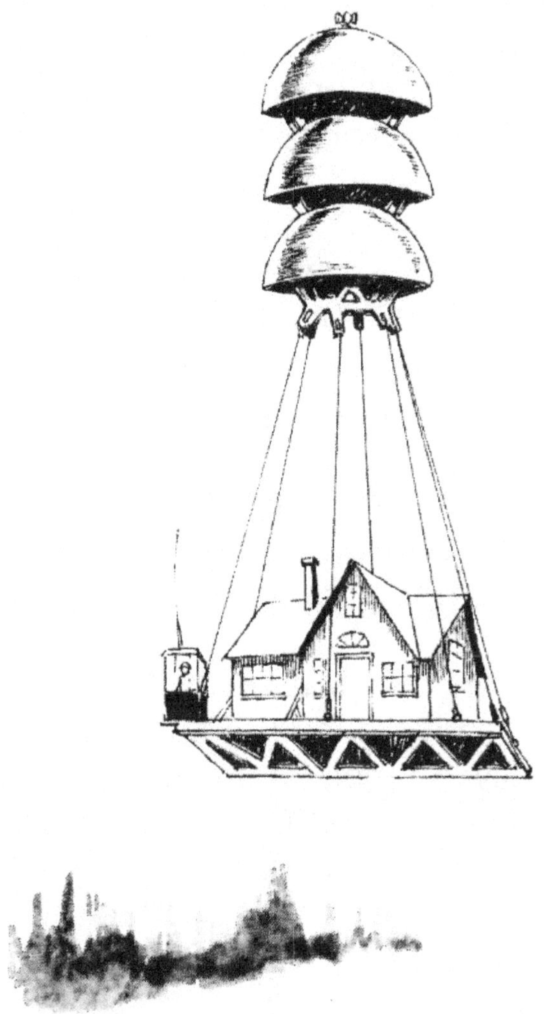

Fig. 2-9

House moving made easy.

Putting a payload into earth's orbit would be a fraction of the cost of contemporary technology. With the use of the reactionless drive in space we may begin colonizing other worlds within a generation. Travel near (or possibly beyond?) the speed of light makes the vast, incomprehensible distances between stars navigable (Fig. 2-10).

Access to the solar system and the stars would open up countless positions in exploration, research, and mining; we may find more gold and diamonds out there than we'll know what to do with! Vital natural resources could become abundant. The availability of unlimited territory should ease pressure on territorial disputes and land acquisition. Exotic jobs would come into being: How would you like to be a tour guide on excursion trips around the solar system?

When the curtains of the reactionless drive-based future have been fully opened there will be no losers. New opportunities and frontiers, hitherto inaccessible, could have deeply inspirational effects on humankind. Every single one of us will benefit.

Fig. 2-10

The speed potential is whatever the ultimate universal limit is. At present, this is believed to be the speed of light.

Fig. 2-11

Transport goods quickly, efficiently, and quietly.

3

THE CONTROVERSY

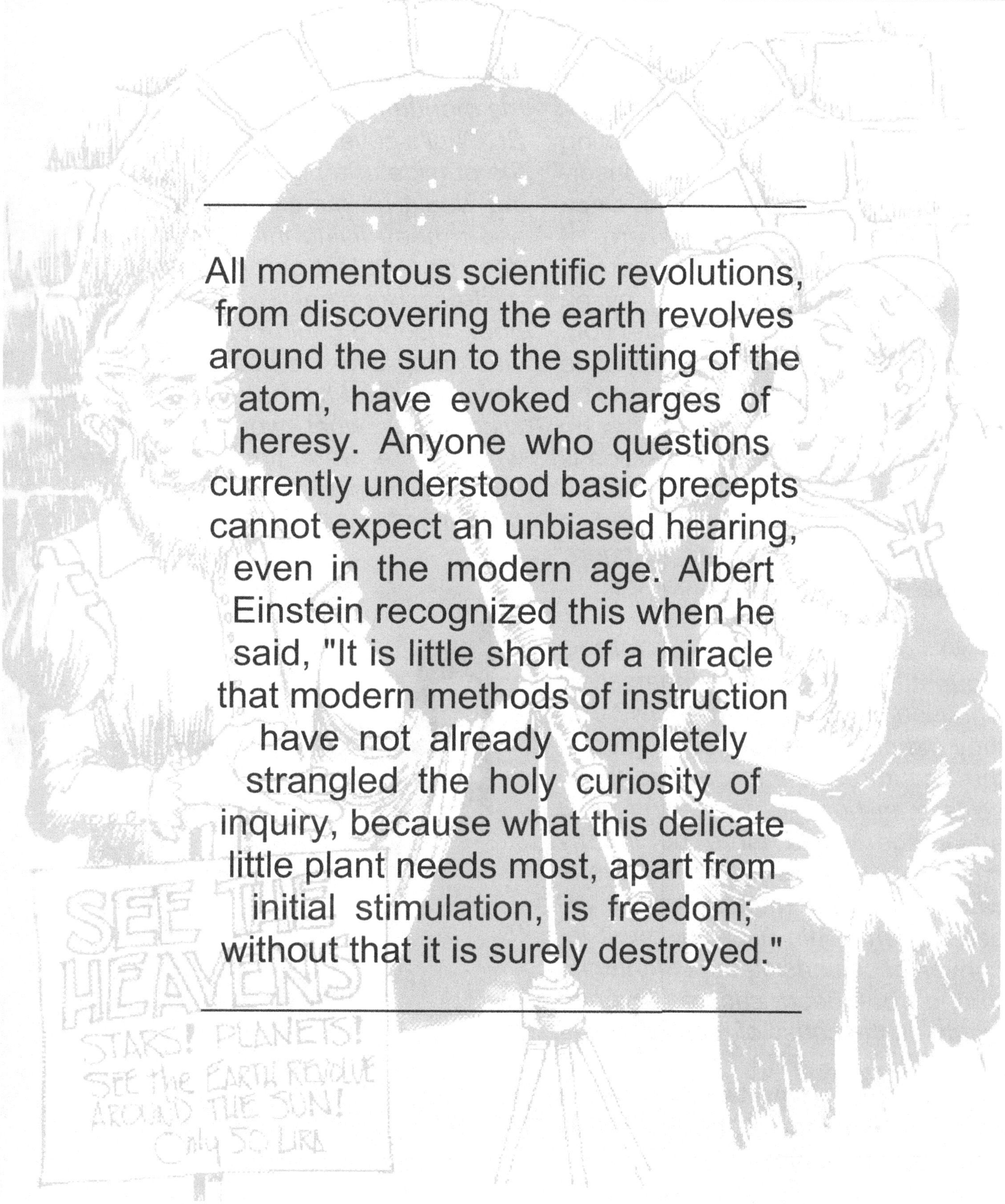

All momentous scientific revolutions, from discovering the earth revolves around the sun to the splitting of the atom, have evoked charges of heresy. Anyone who questions currently understood basic precepts cannot expect an unbiased hearing, even in the modern age. Albert Einstein recognized this when he said, "It is little short of a miracle that modern methods of instruction have not already completely strangled the holy curiosity of inquiry, because what this delicate little plant needs most, apart from initial stimulation, is freedom; without that it is surely destroyed."

Examples of this abound.

A century or so ago many sincerely believed that it would be impossible to do work with alternating current without violating the conservation of momentum principle, since the average current was zero. It turned out that although the current in one direction is balanced by an equal flow in the opposite direction, the flows are not equal and opposite simultaneously and thus work *can* be done.

The whole history of aviation is also full of those who scoffed at the thought of people flying. In 1685 a book called the *Morass Diggers' Jocosa* remarked with contempt:

"How often have those fellows brought ridicule upon themselves who sought to make men fly? For they did not know that it is written in the Scriptures, 'As the bird to flight, so is man born to work,' and what advantage would there be in any case if men could fly? . . . Anyone who looks at the matter properly will see not only little use, but great inconveniences arising out of fluttering to and fro. Indeed, such an inventor would be the cause of many deaths."

In 1713 Joseph Addison warned a would-be inventor of the *immorality* flight would enable:

". . . A couple of lovers would make midnight assignation upon the top of the monument . . . See the Cupola of St. Paul's covered in both sexes like the outside of a pigeon house. Nothing would be more frequent than to see a beau flying into a garret window, or a gallant giving chase to his mistress, like a hawk after a lark. The poor husband could not dream what was doing over his head: If he were jealous indeed he may clip his wife's wings, but what would this avail when there were flocks of whoremasters perpetually hovering over his house?"

"Hawk hunting a lark."

French savant Joseph Lalande declared in 1782:

"It is entirely impossible for man to rise into the air and float there. For this you would need wings of tremendous dimensions and they would have to be moved at three feet per second. Only a fool would expect such a thing to be realized."

Yet scarcely a year later a hot air balloon finally lifted off from the ground—and what a spectacle that was! Half of the population of Paris gathered to watch this first public demonstration despite a downpour. The quietly floating, unmanned balloon rose three thousand feet, vanished into a cloud, reappeared again, and gradually drifted away. After two hours aloft its gas envelope split and the balloon descended upon an unsuspecting village fifteen miles away.

In the village pandemonium ensued. Many fled, while a few bold citizens approached the thing that moved and smelled of sulfur, which two monks confirmed was a monster from Hell. It was then attacked with stones, pitchforks, and firearms. After riddling the balloon with holes, these brave villagers "tied the tool of the finest physical experiment that ever had been made to the tail of a horse, and dragged it a thousand fathoms across the field."

No one took the first successful airplane much more seriously. After decades of reported "flights" and failures by scientists, half-baked inventors, thrillseekers, and cranks, the press was wary. Few in 1903 believed the brief announcement in the paper that Orville and Wilbur Wright, two obscure bicycle mechanics from Ohio, had completed the first powered, heavier than air flight. After all the sound and fury that had surrounded flying for centuries, there was dead silence when success was at last attained. Several years of frustration followed before their amazing accomplishment was acknowledged.

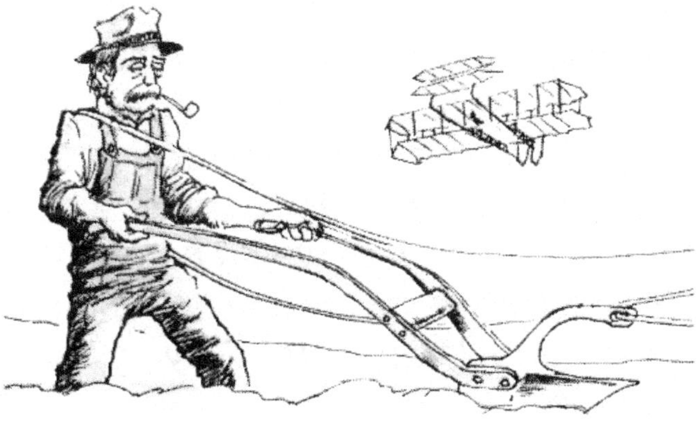

"And to think the world's experts are still telling the Wright brothers they can't fly."

Dr. Robert Goddard, inventor of the liquid fueled rocket, encountered negative reactions in the 1920's when he proposed that his invention could allow travel to the moon. The *New York Times* lambasted him, saying:

"As a method of sending a missile to the higher and even to the highest parts of the Earth's atmospheric envelope, Professor Goddard's rocket is a practicable and, therefore, promising device . . . It is when one considers the multiple-charge rocket as a traveler to the Moon that one begins to doubt . . . for after the rocket quits our air and really starts on its longer journey, its flight would be neither accelerated nor maintained by the explosion of the charges it then might have left. That Professor Goddard with his 'chair' in Clark College and the countenance of the Smithsonian Institution does not know the relation of action to reaction, and the need to have something better than a vacuum against which to react, to say that would be absurd. Of course he only seems to lack the knowledge ladled out daily in high schools."

It is easy to mock revolutionary ideas, but eventually the truth comes to light. People should not take offense at the concept of the reactionless drive. The existence of an internal force that influences motion has been proven. Several working models demonstrating this fact have been built for which patents have been issued. The purpose here is not to discredit science but to advance knowledge and thus improve the living standards of the human race. Researchers and scientists should participate in developing this idea.

Carl Sagan noted, "The thing Galileo fought for—-the things for which science has honored him, classified him as a martyr for science—was the fundamental proposition that demonstration must be accepted; that observational data must never be suppressed for the sake of authority and theory."

John W. Campbell, Jr., the editor of *Analog Science Fact and Fiction* magazine, devoted almost an entire issue of his magazine to championing the quest for the reactionless drive. The following article echoes many of our own views.

Galileo being corrected by the experts of the day.

THE SPACE DRIVE PROBLEM
(*Analog,* June 1960)

It may seem at first thought that the problem of a space drive is a purely technical problem. It seems clear enough that if we want a mechanism, or principle, by which a vehicle can be propelled in the free-space—a device not a rocket, but something acting on the level of force fields, that does not have to carry reaction-mass to throw away—this is a pure, physical-science problem.

It isn't. It's a violently emotional problem, first, a redhot political problem second, and only incidentally a technical problem. Basically, the technical problem is the easiest of the three.

The reasoning behind that statement is quite simple; Nature invariably gives without fear, prejudice, or dishonesty. The technical problem is simply that of asking Nature the right question.

The other two aspects of the problem do not have the same clear-cut simplicity. Both involve human emotions—which, as various philosophers have reported over the last six millennia of recorded history, are anything but clear-cut or simple.

In the first place. the most honorable and ethical of man can be a bald faced liar, if he's misinformed himself. Even a man so inhumanly honest as to be able to overcome completely any personal emotional bias can still be misinformed.

If you think that there are no emotional problems entailed in the space drive problem . . . please think again, including more of the relevant facts. Is it an unemotional problem to a man who has devoted fifteen years to rocket-engine research and development? To an executive who has been responsible for authorizing the expenditure of hundreds of millions of the national wealth on the development of launching-pad facilities?

The buggy-whip manufacturers didn't believe, when the Model T appeared, that their industry was finished. The fact dawned on them only slowly. But gradually they did come to realize that there was no possible improvement to buggy-whip design that could, by brilliant superiority, regain the dwindling market. It wasn't a matter of competition with their product; it was the horse—without which the buggy-whips had no meaning—that was innately incompetent to compete.

There is no possible brilliant improvement in rocket design that can make it competitive with a true space drive. The fact is perfectly, and unarguably clear to any rocket engineer. Unlike the buggy-whip manufacturer, who only slowly came to realize that his industry no longer existed, the rocket engineer can see at once that rockets are reduced to a very small-time hobby or special-effects business. If you want to drill a hole a few inches in diameter through one hundred feet of hard rock, a rocket—the double-ended type—is by far the simplest, cheapest, most probable and quickest technique.

Who wants a true space drive, then? Not the rocket engineers! And not the scientists in general—not when it means the destruction of the foundations of their science. If one can't rely on the eternal validity of Newton's laws of motion . . . what stability is there in the world of Science? It's not just a space drive; it's a thing that casts doubt on the validity of the laws of fluid flow, the conservation of energy, the laws of thermodynamics— on everything!

Because to be a space drive—not antigravity, which isn't a drive, but simply something that takes off the parking brake, so to speak—the device must, in some fashion, negate the Newtonian Laws of Motion. It can't drive in space without drastically rearranging the law of conservation of momentum, and the law of action-and-reaction. And anything that leaks through the law of conservation of momentum automatically challenges the law of conservation of energy. The laws of thermodynamics are based solidly on those; invalidate, or even seriously challenge them, and thermodynamics is a structure without a foundation.

Relativity is based solidly on the conservation of momentum, mass-energy, and electric charge. Any true space drive throws two of the three in doubt. This is something to make a scientist feel happy and contented?

But that no government agency either accepted a demonstration, or bothered to inspect the device, until after the patent was published, and it had been discussed in the December, 1959 editorial, is not opinion. It's checkable fact.

The scientists of the National Aeronautics and Space Administration specifically violated that fundamental for which Galileo fought. They wouldn't look. Neither would the Office of Naval Research. Neither did anyone from the Senate Space Committee. Which is perhaps more remarkable; a Senate committee that rejected an opportunity to investigate something! . . .

X-rays, electronics, cyclotrons, and electron microscopy all stemmed from Faraday's development of the generator, the transformer, and the motor. They have been the great new-fields application of electric power....

The principle of operation is, of course, what's in dispute. Science holds the device to be a "non-member of a non-existent class"—a non-existent class of "devices that don't conform to the law of conservation of momentum."

When Newton did his work, he had, buried under it all, an unstated, and unanalyzed assumption; that there was, of course, one, and only one possible frame of reference.

The whole of Newtonian and classical physics rested on that assumption; it worked fine until toward the end of the nineteenth century; in the beginning of the twentieth century it was really in trouble.

Einstein correctly spotted, and challenged the assumption, and showed how to handle many unresolvable problems, in terms of multiple frames of reference. But . . . with one underlying catch. Einstein had no mathematical tools competent to analyze more than one relationship at a time; therefore he was forced to simplify the problem of reality by saying "there is not simultaneity."

The essence of the situation is—whether modern orthodox physics likes it or not—that our laws of energy and momentum are, in fact very special cases of much more general realities. Newton we already know was fundamentally in error; it is essential, in cosmological physics, to consider more than one frame of reference. Einstein demonstrated that.

But since our laws of conservation stem from Newtonian concepts—they are suspect anyway, and they were before [Norman] Dean's devices came along.

Dean's proposal was rejected on the grounds of pure theoretical consideration. That's the same grounds on which the Church Fathers rejected Galileo's proposals, and refused to look through his telescope.

It might be helpful if all science students were required to study, as part of their college indoctrination, the papers of Galileo and the Church Fathers who were kindly, but firmly, correcting him. Also some of the choicer bits of Newton's and Hooke's remarks concerning the mental competence of their opposers.

Washington is a fascinating state of mind; it operates purely on the pain-avoidance drive. A bureaucrat who does exactly what his directives specifically require, and absolutely nothing else—neither more nor less—avoids the pain of being fired. You can't fire him for failure to accomplish what might have been done; you can fire him only for not doing what his orders require.

It's very rare indeed that someone in government can stick his neck out, and achieve something over and above his assignment. The last notable instance was Admiral Rickover's remarkable achievement of forcing the Navy into nuclear propulsion. As is now well known, he very nearly had his career crushed by the high-brass opposition; Congress saved his bacon, not a "grateful" Navy.

The scientists, in Edison's day, had mathematical proof that the maximum possible efficiency of an electrical generator was fifty percent. They still had the mathematics after Edison started manufacturing ninety-eight percent efficient generators.

4

THE INVENTOR: BOB COOK

Youngest in a family of eight children, Bob Cook, the inventor of the Cook Coriolis drive and the Cook Inertial Propulsion engine, was born in Presidio, Texas on March 1, 1934. His parents were Fred Cook (world-traveling geologist and mining engineer) and Jesusita Rodriguez Leaton (great-granddaughter of Ben Leaton, a legendary hero of southwest Texas).

His family soon realized that Bob was no ordinary child. His insatiable curiosity about nature and how things worked was evident early on. At the age of 5, he built a flashlight with a discarded Hershey chocolate can, two batteries, and a light bulb. He then educated himself on the subject of running wires from electrical sources and was soon able to illuminate his playhouse.

From left to right: Tom Cook, Ann Cook, and Bob Cook. A 1942 photograph.

Anything that spun fascinated the boy: the spin of a gas-driven or electric motor or, in nature, the spin of whirlpools and whirlwinds. To satisfy his curiosity, he ran inside dust devils to see what they felt like and how they worked. He built small windmills to observe them spin in the breeze.

He and his family moved to Nevada. Having been raised mostly mostly by his mother and Spanish-speaking aunts in a Texas border town, Bob only knew a few words of English and had to learn the language.

Bob's love and concern for human life was extraordinary and in his childhood these sentiments went far deeper than anyone would have expected. When he was 11, he witnessed horrifying newsreel footage of a Nazi concentration camp, with scenes of dead bodies stacked like firewood. His shock was so profound that his voice took a sorrowful tone, he would not eat for days on end, and he could not laugh for almost a year. In time he regained his normal voice and the ability to laugh, but the psychic scar never fully vanished.

Eventually the Cooks settled in Concord, California. As a teenager, Bob earned pocket money repairing automobiles. He was especially adept at fixing certain European models that local mechanics had difficulty troubleshooting. He did not do well academically in high school, though, since he found it boring.

His parents had hopes their son would complete a formal education, but the young man rapidly experienced the same boredom with his studies while in junior college and soon left school for his real love—the world of machinery.

At age 19 he was hired as a printing press apprentice for the *Walnut Kernel* of Walnut Creek, California. When, a few months later, the head pressman quit without notice and there was no one with experience to run the press, Bob persuaded his shop foreman to let him try. That night this inexperienced apprentice ran off 20,000 papers, returning the next day to run off another 12,000, leaving his boss who saw it happen astonished. Within a few weeks Bob shattered all production records on that old press with a quality of printing second to none in the area. An article in the *Concord Journal* in 1969 remarked, "It was the fastest apprenticeship ever served. In a matter of a few weeks, he was able to master the problems of a fairly complicated newspaper press, that usually takes an apprentice five years."

Bob worked for various publishers in the printing trade for the next 17 years. During that span he achieved recognition as a mechanical genius. He was responsible for at least seven different inventions in his field, as well as several innovations for operating the presses.

While Bob worked at the *Walnut Kernel,* an important advertising account demanded an extra color for his ads. To make the press print an extra color required installing a special color attachment. The factory had two different attachments available, one for around $5,000 and another for $7,500.

Before purchasing one, Lyman Stoddard, Jr., shop foreman and son of the editor of the newspaper, arranged to rent another press which already had the color attachment installed. Bob ran the color ad on the rented press, but was never satisfied with it. He thought about the situation and proposed to Stoddard that he, Bob, could build a customized color attachment for the *Kernel* press.

Stoddard initially discouraged this, but Bob persisted until the foreman agreed to let him build it for the absurd sum of only $500 for necessary parts and labor.

Left to right: Bob at 21; his godmother Gavinita Spencer; mother Jesusita Cook; and sister-in-law Donna Cook.

As Bob evaluated ways to assemble the attachment, Stoddard suddenly shortened his deadline due to another job that also required an extra color. Although he agreed to the now crushing timeframe, Bob was forced to redesign his idea to make it less complicated. He soon simplified it so much that he was able to put the color attachment together in four hours, with a shocking total parts cost of only 26 cents!

Many who first saw it scoffed at this "flimsy" gadget and thought Bob was crazy to believe it would work. They also thought Stoddard was foolish to risk losing such a valuable advertising account by putting faith in the device. If something so simple could work, they reasoned, then why had the multi-million dollar printing press factory not done the same thing 50 years ago when the press was first designed? Bob knew why—the designers had just never thought of it.

In contrast to this skepticism, Bob's attachment proved to work far more precisely and was far easier to operate than the factory-built models.

Stoddard was proud of Bob. In 1980 he would recall Bob as being "by far" the best pressman he had ever seen in the business. Stoddard laughed when he remembered the old stuffed leather chair Bob positioned near the press. When the press was fully adjusted, Bob would relax in his chair and close his eyes. It was unbelievable. No one could sleep and run that press—it was one of the worst in the business! Yet Bob was producing better results than could be obtained from the factory specifications of many newer model printing presses.

What did Bob actually do in that chair? Sleep? Daydream? Stoddard never knew. Years later he guessed that maybe running the printing presses had become too easy for his employee, whose thoughts probably wandered to "perpetual motion or other impossible dreams."

"You see that guy over there? If I ever catch you doing that on the job, you're fired."

that maybe running printing presses had become just too easy for his young employee, whose thoughts probably wandered to "perpetual motion or other impossible dreams."

During this time, Bob was rapidly becoming an expert in rotary motion, gaining practical experience and an on-the-job education. He learned to consider the critical factors involved in surface speeds of spinning masses and how to transfer an object from one set of rollers to another at very high speeds. He also learned the incredibly complex gearing combinations needed needed to run multiple units at different time phases.

Most importantly, he developed deep understanding of forces and inertia. Tremendous centrifugal and Coriolis forces are developed in printing presses, which have huge rolls of paper and heavy lead printing plates spinning at high speed, with blades continuously folding in and out of the main cutting cylinder. All this knowledge, combined with his acute intuition, provided the necessary tools for developing new, spin-based propulsion systems.

LIVERMORE VALLEY PUBLISHING CO.

LOWELL E. JESSEN, Publisher
M. R. HENRY, Associate Publisher

59 So. Livermore Ave. — P. O. Box 31 — Phones: HIlltop 7-2111, HIlltop 7-2112, HIlltop 7-3545

LIVERMORE, CALIFORNIA

March 11, 1960

Dear Bob,

Again my personal thanks for helping us over several tough problems on our press runs.

In fact, you're the Babe Ruth of the flat bed web circuit.

We appreciate your willing and understanding disposition, and trust that our paths will cross again.

Hope you have a lot of fun working on your book, and that you'll enjoy a quiet time among those noisy Texans.

Sincerely,

Lowell Jessen

Lowell Jessen

PUBLISHERS OF —
The Livermore Herald
The Livermore News
Valley Shopping News
Twin Valleys Advertiser

COMMERCIAL PRINTING —
Herald - News Press

A complimentary letter to Bob after he helped out in several emergencies at the *Livermore News* pressroom (Bob worked elsewhere at the time).

April 24, 1958

Mr. Frank Baker, manager
Flatbed Service Department
Goss Printing Press Company
5601 West 31st Street
Chicago 50, Illinois

Dear Frank:

In response to your latest letter, our pressman is preparing to send you drawings and other pertinent data on his applied-for-patent automatic depression compensator.

His name is Robert L. Cook. His address: 200 Waltham Road, Concord, California. Phone: MUlberry 2-4109.

The idea back of it is simple. The compensator itself looks complicated in the drawings, though there is only one difficult pin to perfect.

Bob is prolific with ideas. As I said before, he has an automatic mailing machine that will handle flatbed or cylinder mailings right from the press. One of them is in operation. He worked out a two-color arrangement for a Model E, enabling the shop to print many thousands of spot color handbills and tabloids for large concerns. He is the one who figured how to spot color on our three-decker and had the patience and skill to do it.

Bob's attorney advises him to sell his idea, or take royalties, thus freeing him to work on other ideas. You'll find him a fine person to work with.

Now, when do we get the slitter I ordered from you? We'll need it early in June.

Sincerely,

Lowell E. Jessen

An interesting letter to the Goss Printing Press Company regarding Bob's inventive abilities.

An old Goss printing press similar to the first one
Bob learned to run.

5

THE FAILURES

The concept of trying to propel by producing an internal force is not new. This principle, though declared in violation of the laws of motion, has nonetheless been attempted by inventors worldwide. Over a hundred patents for such devices have been granted.

We have studied most of these patents, adding to our knowledge of inertial forces by analyzing what was tried before. It became clear that to put together a true, working reactionless drive is to traverse a very narrow trail strewn with land mines and booby traps. We have stepped on our fair share.

Over 50 patents had been granted to inventors experimenting in this field through the 1970's, with dozens more following. This list of patents indicates the scope and approaches of these efforts.

1 Atto, Y. *Propulseur magnetique* (4/15/1957). French Patent No. 1,143,489.

2 Auweele, A.J. *Prime mover* (2/3/1970). United States Patent No. 3,492,881.

3 Bahnson, A.H., Jr. *Electrical thrust producing apparatus* (11/1/1960). United States Patent No. 2,958,790.

4 di Bella, A. *Apparatus for imparting motion to a body* (10/8/1968). United States Patent No. 3,404,854.

5 Benjamin, P.M. *Centrifugal thrust motor* (8/7/1973). United States Patent No. 3,750,484.

6 Benson, E.H. *Inertia engine* (2/4/1975). United States Patent No. 3,863,510.

7 Birck, J. *Propulseur a impulsions* (11/18/1963). French Patent No. 1,347,123.

8 Black, J.W. *Non-linear propulsion and energy conversion system* (2/2/1993). United States Patent No. 5,182,958.

9 Bristow, T.R., Jr. *Method and apparatus for converting rotary motion into lineal motion* (10/20/1992). United States Patent No. 5,156,058.

10 Brown, T.T. *A method of and an apparatus or machine for producing force or motion* (11/15/1928). British Patent No. 300,311.

11 Butka, K. *Propulsion system* (5/5/1992). United States Patent No. 5,111,087.

12 Butka, K. *Propulsion system* (8/2/1994). United States Patent No. 5,334,060.

13 Butka, K. *Propulsion system* (4/25/1995). United States Patent No. 5,410,198.

14 Canot, A.C.C. *Propulsion d'aeronefs ou d'autres vehicules par utilisation de l'energie cinetique* (10/3/1966). French Patent No. 1,458,088.

15 Chernin, M.V. *"Pulsating Inertia propulsion unit"* (9/23/1981). Soviet Union Patent No. 865,689.

16 Chernin, M.V. *"Vehicle Inertial Impulse Propulsion Unit"* (12/23/1982). Soviet Union Patent No. 865,690.

17 Clauser, M.U., et al. *Magnetohydrodynamic control systems* (4/13/1965). United States Patent No. 3,162,398.

18 Codebo, A. *Umlaufschlagflugelsystem* (12/25/1934). German Patent No. 632,908.

19 Coleman, W.J., et al. *Electrostatic propulsion means* (1/1/1963). United States Patent No. 3,071,705.

20 Colla, J. *Mechanical propulsion system* (3/25/1986). United States Patent No. 4,577,520.

21 Conti, D. *Sistema per imprimere il moto alle astronavi e agli aerei stratosferici* (7/28/1958). Italian Patent No. 580,085.

22 Cook, R.L. *Propulsion system* (8/15/1972). United States Patent No. 3,683,707.

23 Cook, R.L. *Device for conversion of centrifugal force to linear force and motion* (12/16/1980). United States Patent No. 4,238,968.

24 Cox, J.E. *Dipolar force field propulsion* (5/12/1987). United States Patent No. 4,663,932.

25 Cox, J.E. *Dipole accelerating means and method* (1/2/1990). United States Patent No. 4,891,600.

26 Cuderman, A. *Centrifugal force propulsion* (3/24/1970). Canadian Patent No. 837,448.

27 Cuff, C. *Device for converting rotary motion into a unidirectional linear motion* (7/13/1976). United States Patent No. 3,968,700.

28 Dean, N.L. *System for converting rotary motion into unidirectional motion* (5/19/1959). United States Patent No. 2,886,976.

29 Dean, N.L. *Variable Oscillation system* (5/11/1963). United States Patent No. 3,182,517.

30 Dehen, F.L. *Apparatus for converting rotarty motion into rectilinear force* (9/7/1982). United States Patent No. 4,347,752.

31 Delroy, M.S. and Delroy, M. *Gyrostat propulsion system* (2/25/1992). United States Patent No. 5,090,260.

32 Detraux, D.C.J., et al. *Dispositif anti-gravitationnel* (11/10/1966). French Patent No. 1,377,261.

33 Drescher, B. *Einrichtung zum Beschleunigen and Abbremsen von Fahrzeugen, insebesondere von Raumfahrzeugen* (6/29/1972). German Patent No. 2,061,914.

34 Dobos, E. *Propulsion apparatus* (3/1/1986). United States Patent No. 4,579,011.

35 Drell, S.D., et al. *Method of and apparatus for effecting electro-mechanical energy interchange in a space vehicle* (2/17/1970). United States Patent No. 3,495,791.

36 Dudley, H.C. *Apparatus for the promotion and control of vehicular flight* (6/25/1963). United States Patent No. 3,095,167.

37 Engelberger, J.F. *Space propulsion system* (4/7/1970). United States Patent No. 3,504,868.

38 Estrade, F. *Device for transforming kinetic energy* (4/30/1974). United States Patent No. 3,807,244.

39 Evard, A.H. *Machine transformatrice d'energie* (9/9/1963). French Patent No. 1,340,196.

40 Farrall, A.W. *Inertial propulsion device* (8/16/1966). United States Patent No. 3,266,233.

41 Fellows, J.W. *Thrust vector machine* (8/18/1988). British Patent No. 2,197,426.

42 Foster, R.E. *Converting rotary motion into unidirectional motion* (4/4/1972). United States Patent No. 3,653,269.

43 Fulop, C. *Flywheel* (12/6/1988). United States Patent No. 4,788,882.

44 Gaberson, H.A. *Vibratory motion* (11/4/1975). United States Patent No. 3,916,704.

45 Goldschmidt, R. *Propulsion for vehicles* (10/14/1924). United States Patent No. 1,511,960.

46 Gozlan, C. *Systeme a propulsion inertielle* (1/10/1986). French Patent No. 2,567,202.

47 Gradecak, V. *Electric aerospace propulsion system* (4/13/1965). United States Patent No. 3,177,654.

48 Hagen, G.E. *Flying apparatus* (2/4/1964). United States Patent No. 3,120,363.

49 Haller, P. *Propulsion apparatus* (4/13/1965). United States Patent No. 3,177,660.

50 Halvorson, E.M., and Schwartz, K. *Vibration driven vehicle* (9/29/1970). United States Patent No. 3,530,617.

51 Harvey, J. *Impulse converter* (8/9/1994). United States Patent No. 5,335,561.

52 Hermann, H. *Autokinetischer Antreib* (3/5/1970). German Patent No. 1,556,820.

53 Hull, H.L., and Joslin, D.E. *Reusable mass propulsion system* (5/24/1994). United States Patent No. 5,313,851.

54 Issacson, J.D. and Navarro, T.L. *Translational force generator* (9/29/1992). United States Patent No. 5,150,626.

55 Keks, E.L. *Propulsive vibrator* (2/7/1958). Australian Patent No. 213,927.

56 Kellogg, M.D., Jr. *Gyroscopic inertial space drive* (8/31/1965). United States Patent No. 3,203,644.

57 Ketsurian, A.V. *"Inertial propulsion machine unit"* (3/7/1983). Soviet Union Patent No. 1,002,707.

58 Kethley, L.I. *Gyroscopic propulsion device* (11/15/1988). United States Patent No. 4,784,006.

59 Kidd, A. *Gyroscopic apparatus* (6/18/1991). United States Patent No. 5,024,112.

60 King, J.F., Jr. *Magnetohydrodynamic propulsion apparatus* (5/30/1967). United States Patent No. 3,322,374.

61 Knap, G. *Orbital propulsion apparatus* (5/2/1978). United States Patent No. 4,087,064.

62 Kuhnen, J. *Getriebe mit Veranderlicher Ubersetzung* (2/13/1926). German Patent No. 425,244.
63 Laroche, A. *Systeme de locomotion sans transmission* (6/19/1923). French Patent No. 559,565.
64 Laskowitz, I.B. *Centrifugal variable thrust mechanism* (4/10/1934). United States Patent No. 1,953,964.
65 Laskowitz, I.B. *Centrifugal variable thrust mechanism* (7/30/1935). United States Patent No. 2,009,780.
66 Lehberger, A.N. *Centrifugal propulsion drive and steering mechanism* (8/5/1975). United States Patent No. 3,897,692.
67 Lieurance, R.L. *Centrifugal inertia drive* (2/6/1996). United States Patent No. 5,488,877.
68 Llamozas, J.D.M. *Direct push propulsion unit* (4/28/1953). United States Patent No. 2,636,340.
69 Lundberg, F.R. *Mechanical propellant and steering machine* (11/21/1963). Australian Patent No. 267,091.
70 Marsh, R.O., Jr. *Centrifugal force drive machine* (2/14/1995). United States Patent No. 5,388,470.
71 Mason, L.M. *Centripetal device for concentrating centrifugal force* (2/12/1991). United States Patent No. 4,991,453.
72 Mast, O. *Propulsion system* (6/17/1975). United States Patent No. 3,889,543.
73 Matyas, L.B. *Propulsion apparatus* (6/15/1971). United States Patent No. 3,584,515.
74 McAlister, R.E. and McAlister, T.J., Jr. *Propulsion system* (9/4/1973). United States Patent No. 3,756,086.
75 McMahon, J.C. *Energy transfer device* (12/1/1992). United States Patent No. 5,167,163.
76 Melnick, H.S. *Unidirectional force generator* (4/14/1981). United States Patent No. 4,261,212.
77 Melnick, H.S. *Continuous force and impact generator* (2/23/1988). United States Patent No. 4,726,241.
78 Miller, C.L. *Reciprocating Mechanism* (11/1/1918). United States Patent No. 1,280,269.
79 Modesti, J.N. *Manned disc-shaped flying craft* (11/3/1970). United States Patent No. 3,537,669.
80 Montalbano, P.J. *Conversion of rotational output to linear force* (8/15/1989). United States Patent No. 4,856,358.
81 Montalbano, P.J. *Conversion of rotational output to linear force* (8/27/1991). United States Patent No. 5,042,313.
82 Motts, B.C. *Airship* (11/6/1990). United States Patent No. 4,967,983.
83 Mundo, J.D. *Universal propulsion powerplant and impulse drive unit for self-propelled vehicles* (9/13/1988). United States Patent No. 4,770,063.
84 Navarro, T.L. *System for generating controllable reference envirionment and steerable translational force, etc.* (12/12/1995). United States Patent No. 5,473,957.
85 Neff, T. *Reaction motor* (7/27/1937). United States Patent No. 2,008,115.
86 Neimann, P. *Vorrichtung zum Ausgleichen von Unregelmassigkeiten, in Gang von langsam laufeuden Maschinen* (9/11/1912). German Patent No. 63,188.
87 North, H. *Apparatus for producing a force* (12/15/1987). United States Patent No. 4,712,439.
88 Novak, L.J. *Centrifugal mechanical device* (5/15/1974). United States Patent No. 3,810,394.
89 Nowlin, A.C. *Device for obtaining directional force from rotary motion* (5/30/1944). United States Patent No. 2,350,248.
90 Okress, E.C. *Quasi-corona-aerodynamic vehicle* (9/2/1969). United States Patent No. 3,464,207.
91 Pages, M.J.J. *Engin pour vols cosmiques* (1/9/1961). French Patent No. 1,253,902.

92 Peltier, J. *Impulseur* (4/5/1950). French Patent No. 967,839.

93 Peppiatt, A.C., et al. *Impulse drive* (6/23/1987). United States Patent No. 4,674,583.

94 Peterson, O.F.A. *Apparatus for producing a directional unit force* (5/17/1988). United States Patent No. 4,744,259.

95 Prevot, C. *Dispositif vibreur a masselottes a double mouvement* (12/12/1953). French Patent No. 1,063,784.

96 Quisling, S. *Propulsion mechanism* (1/14/1930). United States Patent No. 1,743,978.

97 Rakestraw, R.G. *Toy vehicle having resilient supports and self-contained drive means* (7/27/1965). United States Patent No. 3,196,580.

98 Redish, W.L. *Rotary vibrator device* (3/2/1965). Canadian Patent No. 704,568.

99 Reid, A. *Mechanism utilising inertia to obtain translational movement* (3/30/1957). British Patent No. 770,555.

100 Rhodes, C.W. *Improvements in or relating to centrifugal mechanism* (4/18/1921). British Patent No. 162,334.

101 Rice, W.A. *Propulsion system* (10/8/1963). United States Patent No. 3,106,167.

102 Rickman, E.J.C. *Improvements in or relating to thrust producing devices* (7/13/1977). British Patent No. 1,479,450.

103 Rogers, C.E., et al. *System for propulsion and positioning of a transitory object* (1/31/1989). United States Patent No. 4,801,111.

104 Rogers, C.E. *Controllable gyroscopic propulsion apparatus* (10/8/1991). United States Patent No. 5,054,331.

105 de San, M. *Improvements relating to the propulsion of vehicles* (3/23/1960). British Patent No. 830,816.

106 Schieferstein, G.H. *Verfahren und Vorrichtung zur Fortbewegung mechanischer Vorrichtungen* (1/25/1930). Austrian Patent No. 115,928.

107 Schnur, N.J. *Method and apparatus for propelling an object by an unbalanced centrifugal force with continuous motion* (9/14/1976). United States Patent No. 3,979,961.

108 Schur, G.O. *Thrust motor* (3/8/1966). United States Patent No. 3,238,714.

109 de Seversky, A.P. *Tonocraft* (4/28/1964). United States Patent No. 3,130,945.

110 Shimshi, E. *Sphereroll* (6/27/1995). United States Patent No. 5,427,330.

111 Spies, J. *Luft—und/odor Raumfahrzeug* (1/5/1972). German Patent No. 2,032,416.

112 Srogi, L. *Mechanical propulsion system* (1/6/1981). United States Patent No. 4,242,918.

113 Taylor, J.R. *Electromagnetic energy propulsion engine* (3/30/1993). United States Patent No. 5,197,279.

114 Thornson, B.R. *Apparatus for developing a propulsive force* (12/30/1986). United States Patent No. 4,631,971.

115 Trivellin, E. *Dispositivo atto ad imprimere un moto in qualisiasi direzione a dei mezzi mobili* (3/6/1958). Italian Patent No. 573,912.

116 Van Leeuwen, G.H. *"Vehicle Propulsion System"* (12/21/1978). German Patent No. 2,819,409.

117 de Weaver, F., III. *Propulsion system* (10/18/1983). United States Patent No. 4,409,856.

118 de Weaver, F., III. *Propulsion system* (10/30/1984). United States Patent No. 4,479,396.

119 Williams, M.O. *Inertial drive for vehicle has driven rotor* (8/15/1984). British Patent No. 2,097,103.

120 Woltering, H.M. *Rotating eccentric weights vibrator system* (2/14/1995). United States Patent No. 5,388,469.

121 Young, H.W., Jr. *Directional force generator* (1/19/1971). United States Patent No. 3,555,915.

122 Zachyatal, G.J. *Centripetal device for concentrating centrifugal force* (12/5/1989). United States Patent No. 4,884,465.

Any discussion of the various attempts to propel through internal means requires an understanding of some fundamental nomenclature, as well as some knowledge of the basic forces produced by rotary motion. The table at left defines the terms used.

Although it appears that most forces have been accounted for, the following thoughts of Einstein and Infeld should be kept in mind.

"When first studying mechanics one has the impression that everything in this branch of physics is simple, fundamental, and settled for all time. One would hardly suspect the existence of an important clue which no one noticed for three hundred years. The neglected clue is connected with one of the fundamental concepts of mechanics—that of mass."

In like manner, a more complete understanding of inertial forces will unlock several clues concerning the existence of internal propulsive forces that also have gone unnoticed for over three hundred years. Let us briefly examine centrifugal force, centripetal force, Coriolis force, gyroscopic force, and tangential force.

DEFINITIONS OF TERMS USED

angular momentum—determines the energy or force value of the object moving in a circular path. The angular momentum is the factor that determines the energy contained within a spinning mass.

angular velocity—rate of change of direction of a mass.

centrifugal force—the inertial effect produced when an object is constrained to move in a circle. This force can be viewed as the equilibrant of centripetal force.

centripetal force—the inward pulling force that causes an object to move in a circle.

Coriolis force—the inertial effect occurring when a mass is constrained to move radially across a rotating body.

gyroscopic force—resistance to torque that would change the position of the axis of a spinning mass.

impulsive force—force acting for a short time but sufficiently large to cause some change in momentum.

inertia—the tendency for matter to remain in a state of rest or in uniform motion.

kinetic energy—work the object can do by virtue of its motion. The higher the speed, the more work potential.

linear momentum—determines the energy or force value of the object moving in a straight line. The linear momentum is the factor that determines the energy contained within a linear moving mass. Linear momentum is a product of mass and velocity.

linear velocity—speed the object is moving in a straight line.

power—rate of doing work.

rotor—in the Cook system, a propeller-like unit having weights spinning around the center shaft.

torque—twisting or turning action.

CENTRIFUGAL AND CENTRIPETAL FORCE

In our study of the various propulsive mechanisms with spinning masses, we have chosen to view the resulting forces from a rotating frame of reference. The general theory of relativity admits that the explanations of both rotating and non-rotating observers are equally valid. We are concerned with what forces affect the center of the mechanism.

For example, consider a volunteer turning at the same angular speed as a ball attached to the end of a string he is holding (Fig. 5-1). This is a rotating frame of reference and from the volunteer's point of view the ball has no acceleration and is at rest. The outward *centrifugal force* produced by the ball is equalized by the inward pull of *centripetal force.*

In our example, centrifugal force is a very real force for the rotating observer and similarly is an effective force on the center of the mechanisms we will study.

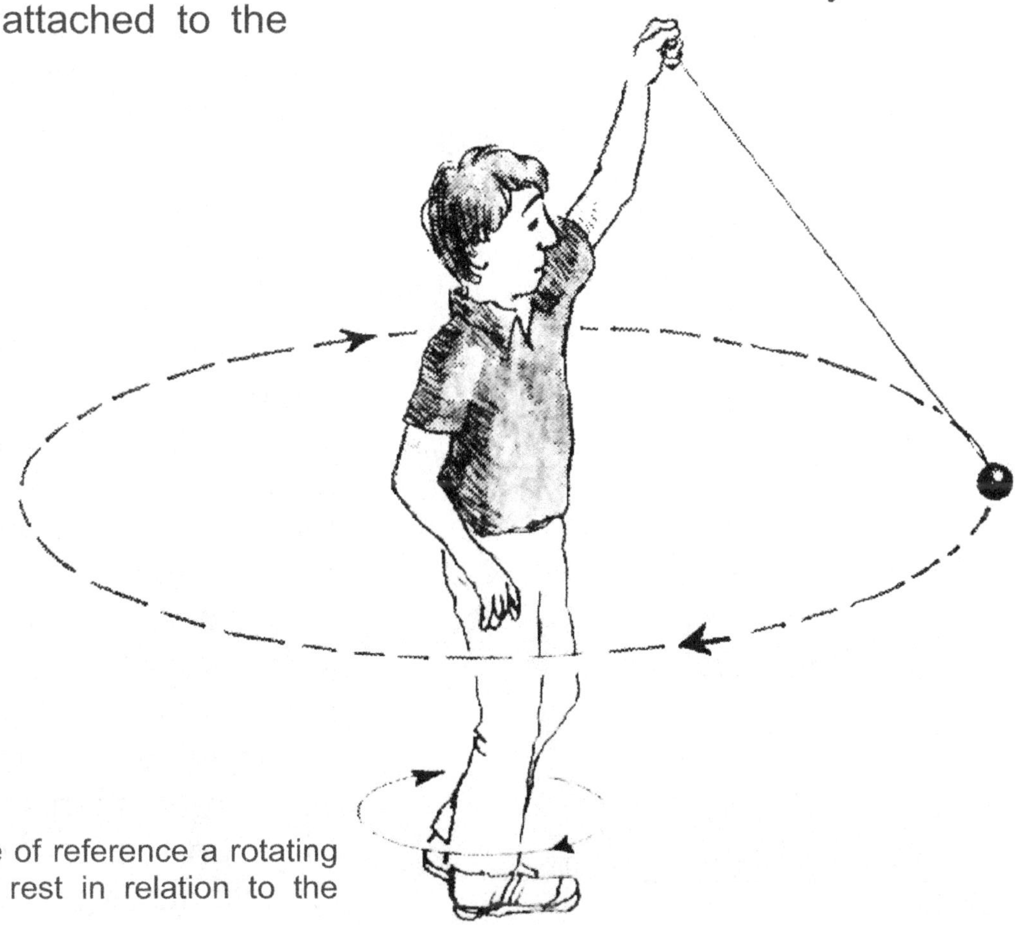

Fig. 5-1

From a rotating frame of reference a rotating ball seems to be at rest in relation to the observer.

CORIOLIS ACCELERATION AND CORIOLIS FORCE

A proper understanding of Coriolis force and acceleration is essential for comprehending many of the patents we will review. The distinction between the two is often confused.

Again, we have chosen a rotating frame of reference. Coriolis force can be defined as *the inertial effect occurring when a mass is constrained to move radially across a rotating body.*

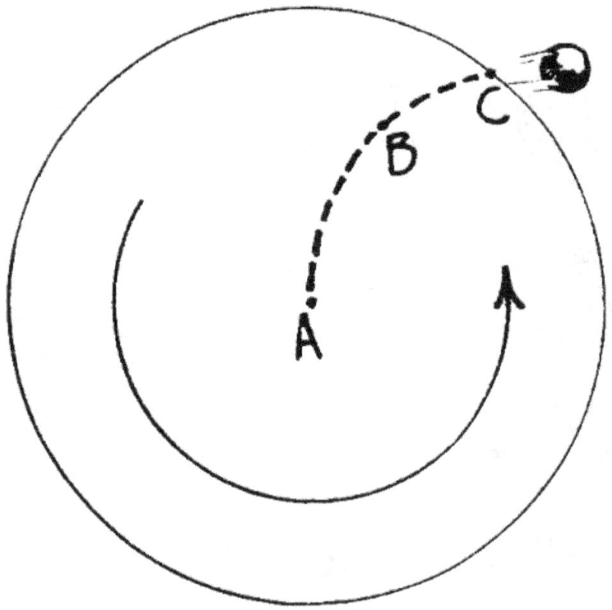

Fig. 5-2

Deflection of ball as seen by rotating observer at center A.

Imagine rolling a steel ball away from center A on a frictionless platform rotating counterclockwise at constant angular velocity. As the ball moves radially away from center A it is unable to match the higher tangential velocity of points B and C. (The ball cannot increase its tangential velocity because there is no friction.) Since the ball does not increase its tangential velocity it *appears* to curve to the right as seen by the rotating observer at center A (Fig. 5-2); an outside observer, however, will see the ball move in a straight line. This is an example of Coriolis acceleration.

Now let us again imagine a steel ball rolling away from center A on the platform rotating counterclockwise. This time, the ball is forced to roll through a smooth tube. The relative acceleration from the previous example now becomes a force pushing on the right side of the tube, trying to slow the angular velocity of the platform (Fig. 5-3). The force is perpendicular to the radial motion of the ball. This negative Coriolis force when pitted against a positive torque then registers a positive force on the center of rotation. The Coriolis force is the same deflecting

A proper understanding of Coriolis force and acceleration is essential for comprehending many of the patents we will review.

force you would feel pushing you sideways should you walk outward on a spinning merry-go-round.

If on the same platform the ball were forced to return from B to A, the opposite would happen. The ball's inertia resists reducing its angular velocity, and it does so by pushing on the left side of the tube, trying to increase the angular velocity of the platform. The force is produced on the left side of the tube (Fig. 5-4).

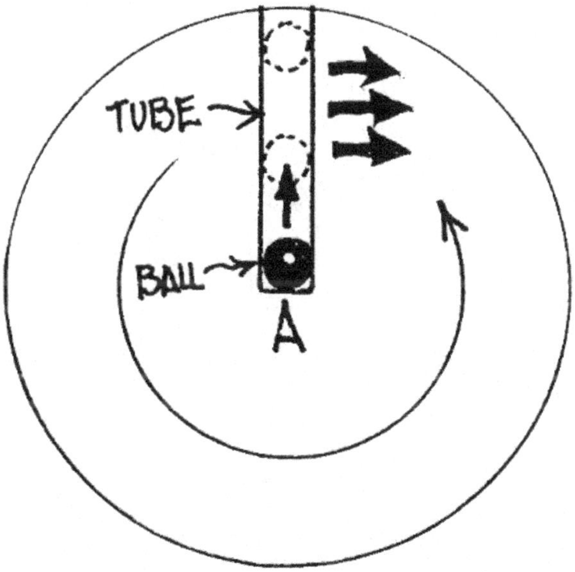

Fig. 5-3

A negative Coriolis force acting perpendicular to the spinning tube.

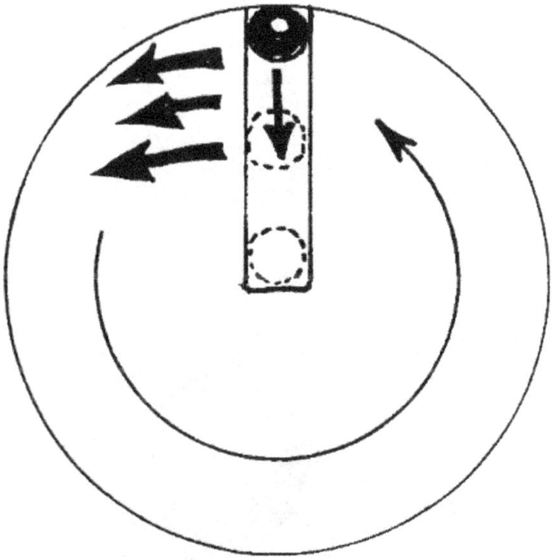

Fig. 5-4

A positive Coriolis force acting perpendicular to the spinning tube.

GYROSCOPIC FORCE

A spinning top is a good example of a gyroscope (Fig. 5-5). The top tends to maintain its position in space because of the inertia of the rapidly spinning body.

A gyroscope rigidly resists being disturbed and reacts to a disturbing torque by precessing (rotating slowly) at right angles to the torque. This principle can be demonstrated with a suspended bicycle wheel spinning at high speed. To observe precession, a force is applied steadily. The wheel is found to precess slowly, not about the axis of the applied torque, but about an axis perpendicular to it and perpendicular to the spin axle (Fig. 5-6).

Any high-speed rotor is like a gyroscope. When torque is applied, the rotor wants to precess. If a rigid rotor is prevented from precessing, a force registers on the system.

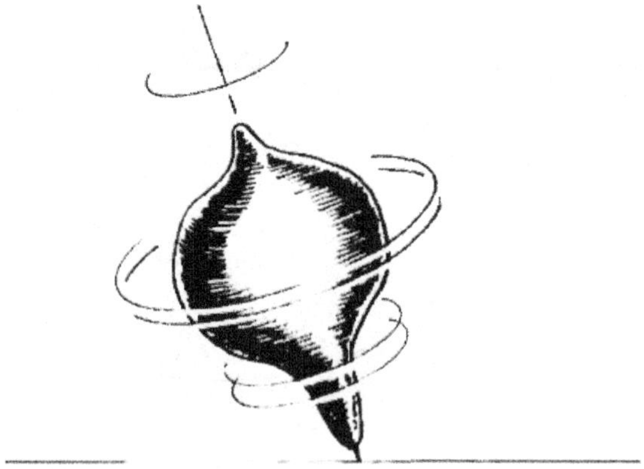

Fig. 5-5

Spinning top.

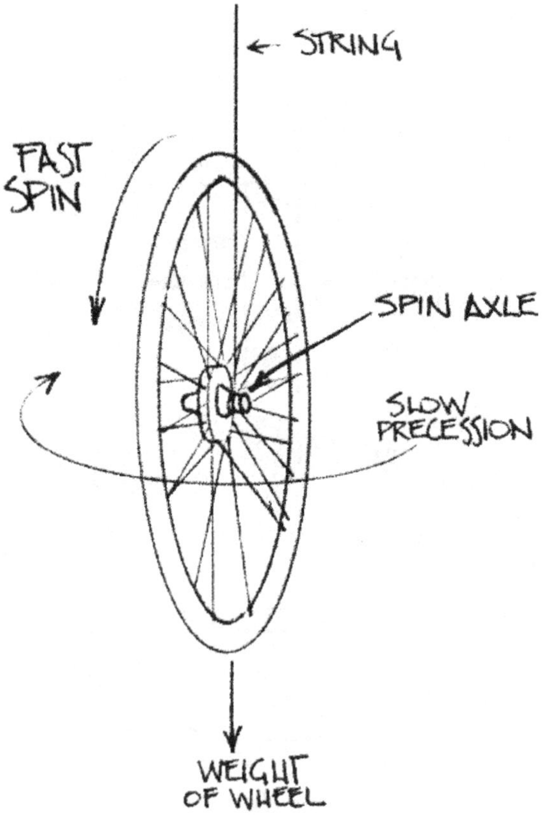

Fig. 5-6

Gyroscopic precession of a spinning bicycle wheel.

TANGENTIAL FORCE

Another force we encountered was a *tangential force*, which is produced when a small spinning mass is briefly angularly accelerated or decelerated. The tangential force can be produced by a brief torque, a change of radius producing a tangential Coriolis force, or by any briefly applied force angularly speeding up or slowing down the spinning mass.

If we briefly change the angular velocity (Fig. 5-7) of the mass at different positions during each revolution and the radius remains constant, outside tangential forces or brief torque will account for the change in angular momentum.

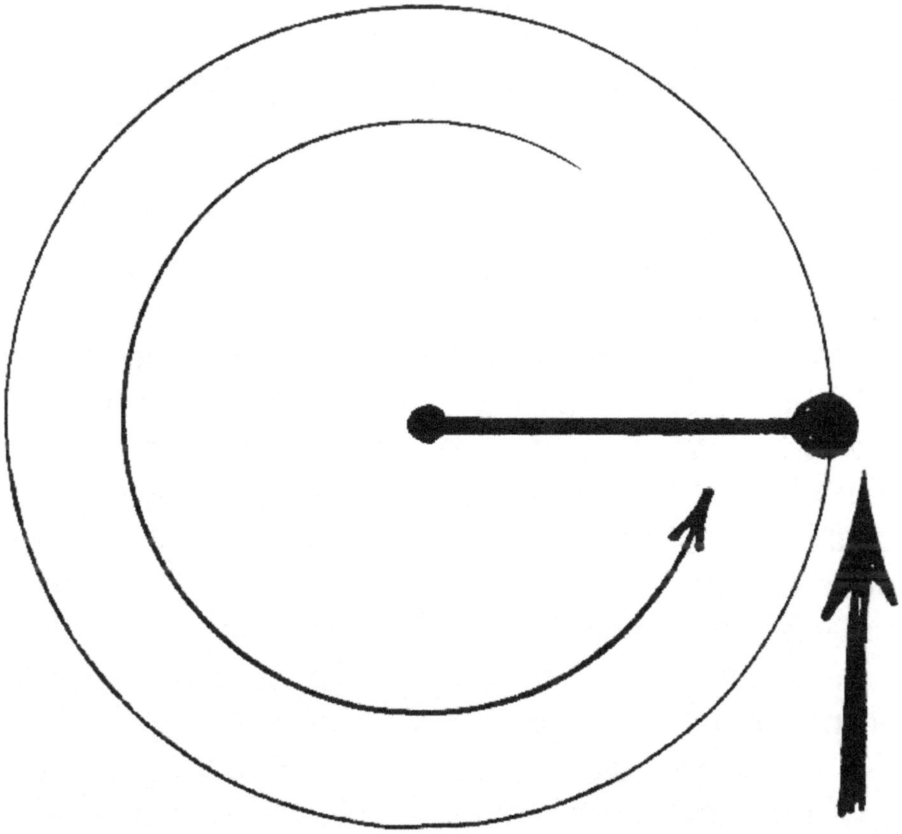

Fig. 5-7

Force vector produced at this point by tangential force or positive torque.

How might we propel using our knowledge of these forces?

Let us imagine a mass spinning in a circle. For 180° of travel from A to B we produce positive centrifugal force, and for 180° from B back to A we produce negative centrifugal force. Over 360°, the forces cancel (Fig. 5-8).

One possibility would be to make our mass spin faster for the positive 180° than for the negative 180° (Fig. 5-9). A greater angular velocity will produce a greater centrifugal force in the forward direction. So let us speed up our mass at point A by applying positive torque, and at point B slow it down by applying negative torque. Unfortunately, this does not work. To speed up our mass at A results in a negative tangential force. The same negative force appears at B when we slow our mass down. The two negative impulses cancel what we gain by the additional centrifugal force. No matter what combinations of acceleration and deceleration we try, we cannot propel with this principle.

Many inventors have tried a slightly different approach to reap the advantages of the powerful and easily generated centrifugal force. Some devices have rotated mass members and shifted the center of gravity relative to the axis of rotation. It thus seems another simple way to propel would be to eliminate the negative 180° of travel.

At position B shoot the mass through the center back to position A, therefore eliminating the 180° of negative centrifugal force (Fig. 5-10). If our mass followed a frictionless path, the lateral force developed at B accelerating the mass through the center could be canceled by the lateral impact force at A stopping its momentum. We know we produce a positive centrifugal force in the forward direction for approximately 180°. Is this, then, an unbalanced force?

Remember, whenever we change the radius of a mass on a rotating body we introduce Coriolis force. At position B the mass is traveling at its maximum angular velocity. When shot toward the center it has to slow down and lose some of its angular momentum. Some of its energy has to be drained. Its inertia at point B resists change, and as it moves radially toward center C a positive Coriolis force results perpendicular to the radial motion, thus producing a negative effect on the system.

From center C back to A the mass now resists increasing its angular velocity (negative Coriolis force), which again produces a negative effect on our center. The two negative effects of the Coriolis forces

5-8

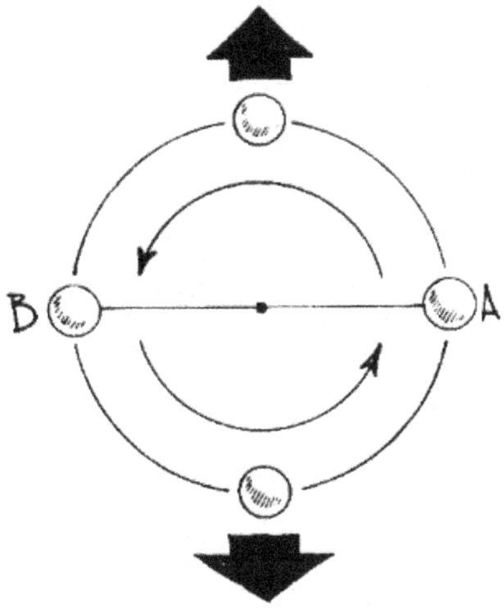

5-9

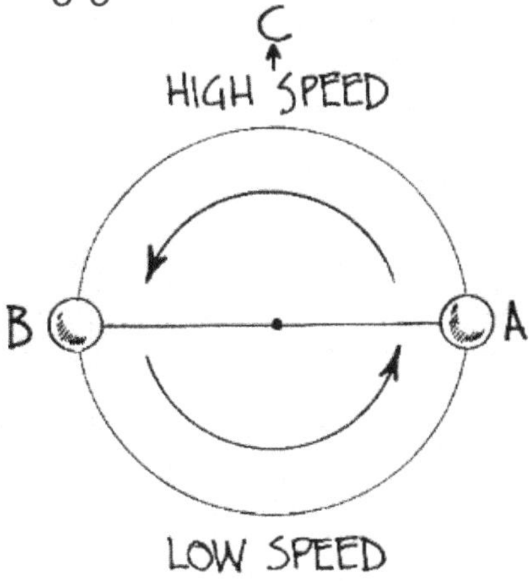

5-10

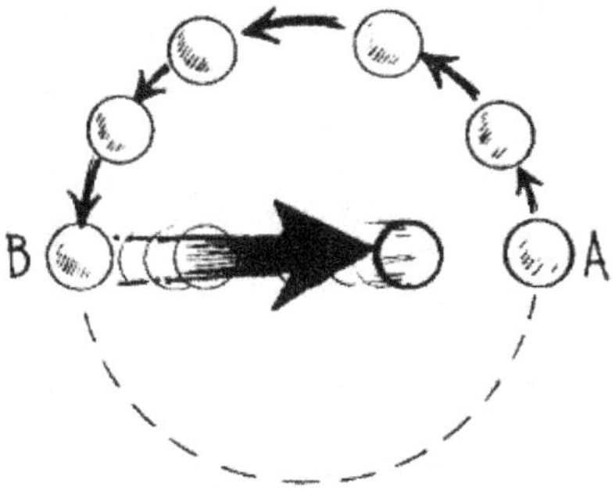

5-11

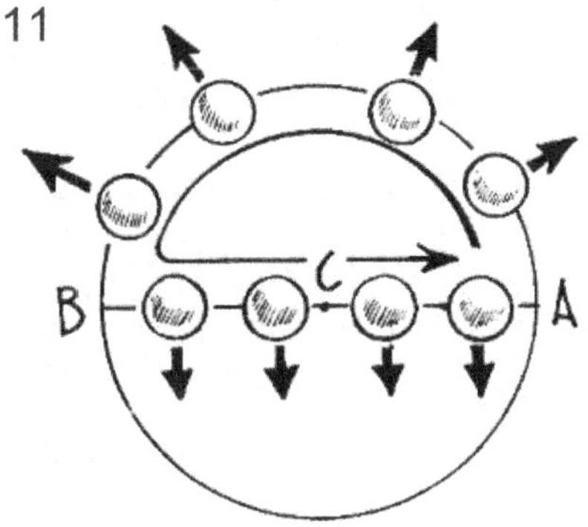

Fig. 5-8

Centrifugal forces in equilibrium.

Fig. 5-9

A positive centrifugal force counterbalanced by negative tangential forces and negative centrifugal force.

Fig 5-10

A proposed method of producing an unbalanced centrifugal force.

Fig. 5-11

Centrifugal forces and Coriolis forces in equilibrium.

have exactly cancelled our positive centrifugal force, and again we have a balanced system (Fig. 5-11).

Several inventors have patented this principle. Witness, for example, the 1934 *Laskowitz drive* (Fig. 5-12). This drive had a series of spinning weights fitted into cylidrical bores. The radius of rotation of the weights would be changed at various points (Mr. Coriolis) to produce a positive centrifugal force that hopefully propelled. The radius of one weight was increased while the radius of the other was decreased.

This motion unfortunately imparted two simultaneous, negatively-acting Coriolis impulses on the center, canceling the positive centrifugal force.

In 1944, the *Nowlin drive* was patented (Fig. 5-13). This drive had a series of gears extending and retracting a series of cranks attached to a propellant mass. The object was to extend the propellant mass in this telescopic fashion at the required position to supposedly produce an unbalanced centrifugal force.

Again, due to both the timing of the mechanism and the shifting position of the weights, the changes of the radius caused negative Coriolis force effects, which canceled the positive centrifugal force.

The *Matyas drive* of 1971 (Fig. 5-14) tried it a bit differently. A pool of mercury was the propellant mass, and the greater concentration of mass was maintained on the positive half of the system by a series of pistons that forced the mercury back toward the center of rotation at certain intervals. This method was yet another way of changing the radius of the propellant mass.

Because of the timing of the machine, the Coriolis again perfectly opposed the lopsided centrifugal forces. Another balanced system.

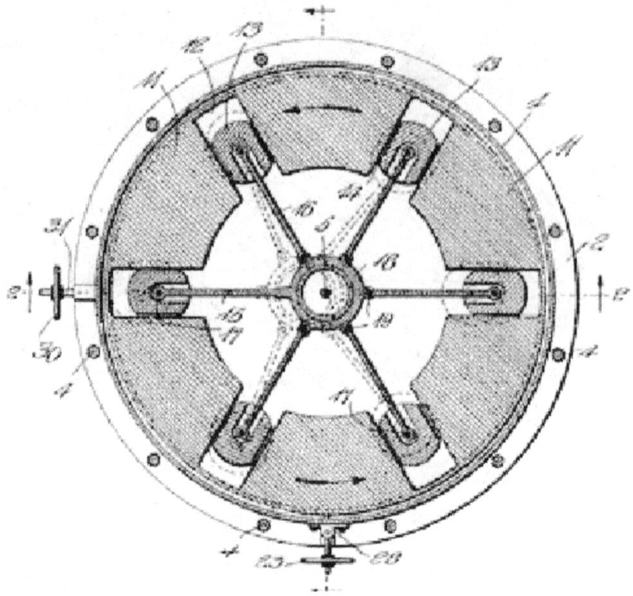

Fig. 5-12

Laskowitz drive.

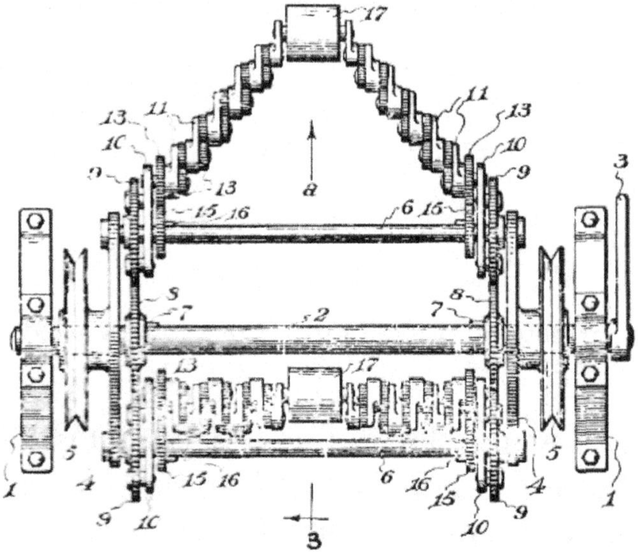

Fig. 5-13

Nowlin drive.

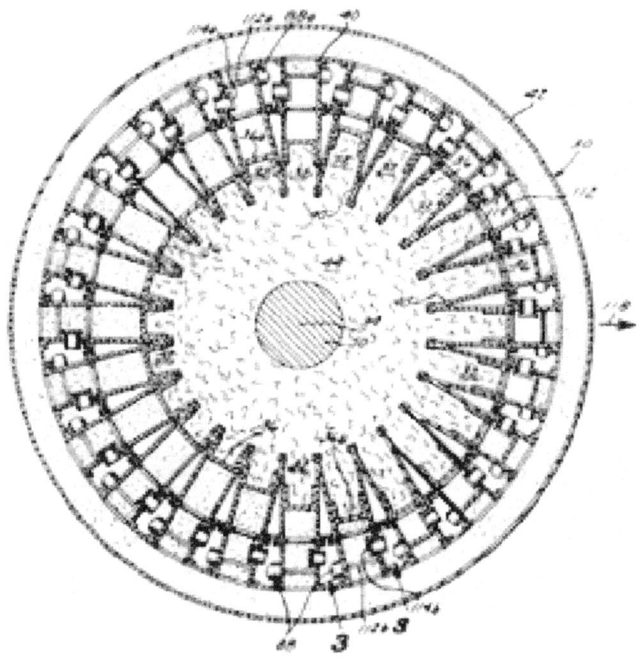

Fig. 5-14

Matyas drive.

The *Novak drive* (Fig. 5-15) was patented in 1974. Novak had a series of off-center rotating masses timed to take advantage of the positive centrifugal force. Same principle, same results—a balanced force.

The *Cuff drive* (Fig. 5-16), patented 1976: Change radius—negative Coriolis effects balance all positive centrifugal force.

There have been yet other patents applying this principle. So far, all have failed because it seems that there is a misunderstanding of the Coriolis force effects. No matter what combination you choose, you cannot produce an unbalanced force and propel quite this way. It seems to be a dictate of nature.

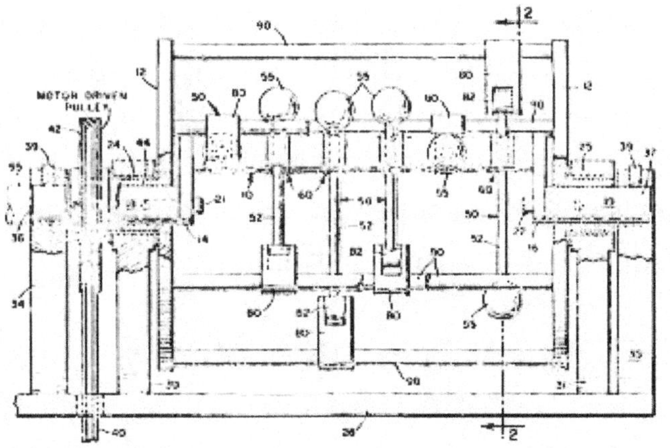

Fig. 5-15

Novak drive.

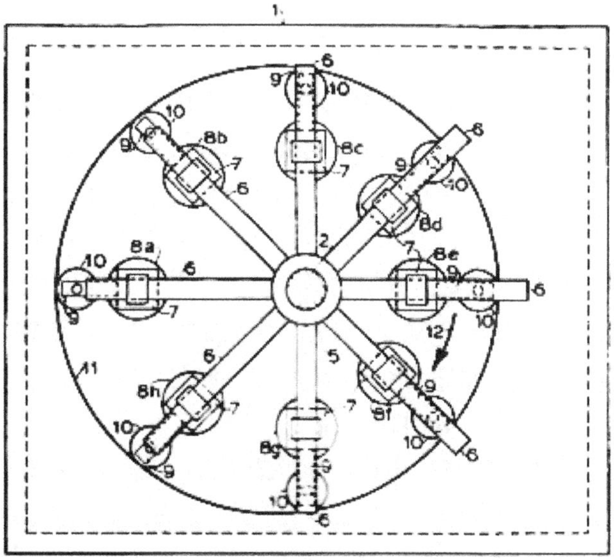

Fig. 5-16

Cuff drive.

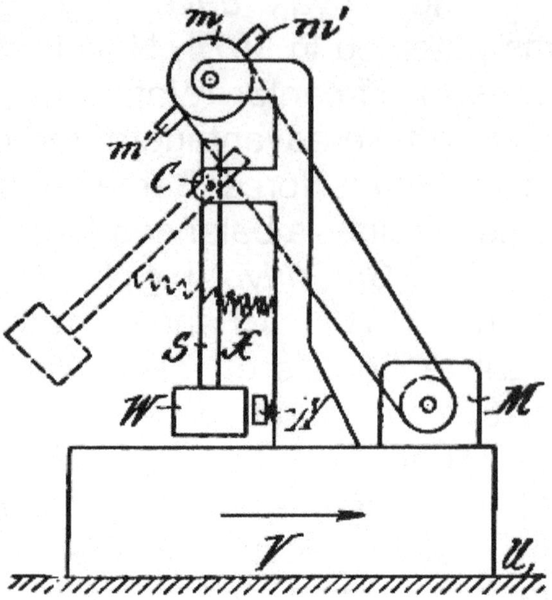

Fig. 5-17

Goldschmidt drive.

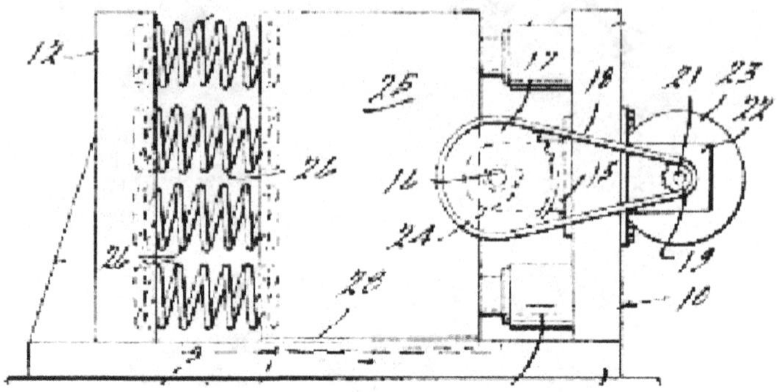

Fig. 5-18

Farrall drive.

Other inventors have tried to propel using a different principle—the sudden application of a force. Known as *impulse drives*, these try to take advantage of a short-lasting powerful force pitted against a longer lasting weaker force with the aid of static friction.

Note the *Goldschmidt drive* of 1924 (Fig. 5-17). A hammer impacted into a stop to impulse the machine forward. This system took advantage of static friction. Friction maintained the machine's position while the hammer was slowly retracted. The release of the hammer produced an impulse overcoming the static friction and moving the machine. Obviously, the machine could impulse forward aided by friction but, in space, would have simply oscillated back and forth.

The *Farrall drive* (Fig. 5-18) was patented in 1966. Here was another battering ram. A large weight compressed springs to a cocked position; then it was released to provide a power impulse. Static friction would again provide limited motion here on earth but bounded motion in space.

One variation of this principle was the widely publicized *Dean drive* (Fig. 5-19), patented in 1959. The mechanism Dean built was rather confusing; it clouded the true picture of the actual mechanical principle involved. The propulsive force in Dean's system was centrifugal force. The oscillating carriage had two counter-rotating weights Dean called "eccentric inertial masses." These produced an intermittent force in the desired direction of travel (Fig. 5-20).

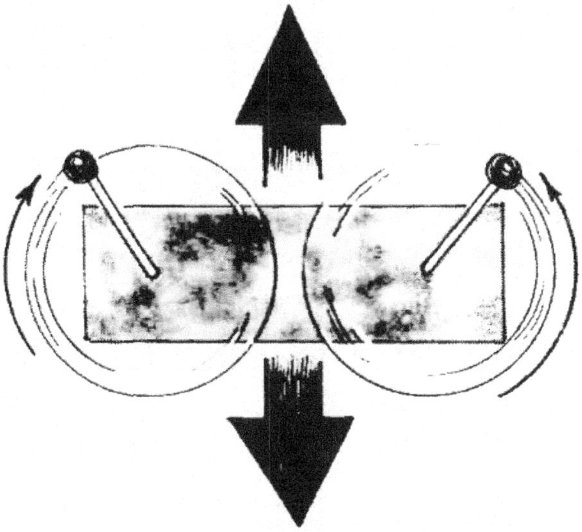

Fig. 5-20

The heart of Dean's mechanism.

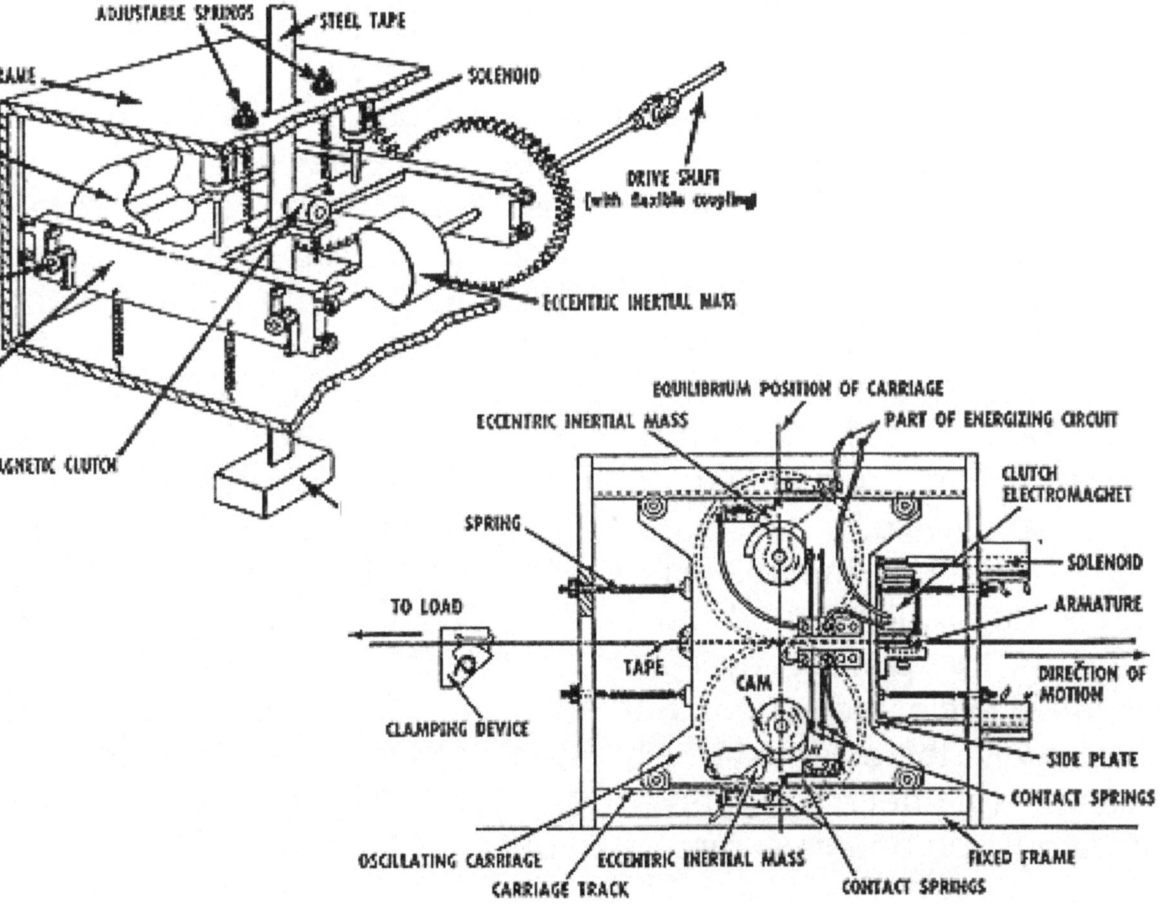

Fig. 5-19

Dean drive.

Dean took advantage of the positive centrifugal force to propel the load, and with the aid of static friction prevented the negative centrifugal force from moving the system backwards. When the weights were producing positive centrifugal force on the load, the electromagnetic clutch grabbed the rigid load tape thus transmitting an impulse that pulled the load forward and moved the carriage into the forward position on the fixed frame.

When the weights swung to the opposite side producing negative centrifugal force, the electromagnetic clutch released the rigid steel tape which was then prevented from moving backwards as the oscillating carriage returned on the track to its starting position. By activating the solenoid at just the right timing and by properly adjusting the springs, Dean could slow the carriage's return and prevent it from overcoming the static friction of the load. (It is important to note that the clamping device provided a rigid connection between the main frame and load on the negative force cycle.)

Therefore, the machine overcame static friction and moved forward on the positive cycle. On the negative cycle, the springs worked in unison with the solenoid which had a cushioning effect and prevented a large enough negative impulse to be translated to the load and move the machine backwards. Dean's machine did propel across the floor. In space, however, without the aid of friction, his model would have only produced bounded motion.

Mr. Jacob Rabinow of Rabinow Engineering did an analysis of a Dean model provided to the Air Force Office of Scientific Research in 1961. One of his tests was to use rollers between the load and its support to minimize friction. With less than one ounce of friction the load oscillated at the same frequency as the carriage, but did not advance toward it. Rabinow concluded that if the frictional forces of the load were smaller than the reverse spring forces, and if the positive and negative impulse were equal, the load would move backward due to spring force, and the net displacement would be zero. This test showed the machine to have no net unidirectional effect on an inertial load if the frictional load is small compared to the mass.

In another test Rabinow used gauges to obtain force-time plots on an oscilloscope. He concluded that both the positive and negative areas were equal and that momentum was indeed conserved.

Rabinow noted that the load moves toward the carriage if: (1) The positive impulse is greater than the load's static friction; and (2) the spring force developed during the remainder of the cycle is less than the static friction. He concluded that the device was incapable of operating as a true space drive.

Static friction aids the impulse drives. Researchers need to find a system that actually benefits from the *lack* of static friction.

More ingenious attempts to propel internally have been patented. One class of drives considers using gyroscopic forces to propel. These ideas are interesting. The *Foster drive* (Fig. 5-21) patented in 1972 and the *Kellogg drive* of 1965 (Fig. 5-22) involved gyroscopic forces. The Foster drive was reported to move across a flat surface at 4 mph. Our experiments lead us to believe that a gyroscopic space drive, although experiencing limited success here on earth, would probably not work efficiently in space. (Some of Bob's experiments with the gyroscopic force have still left him puzzled. He intended to continue these experiments at a later time.)

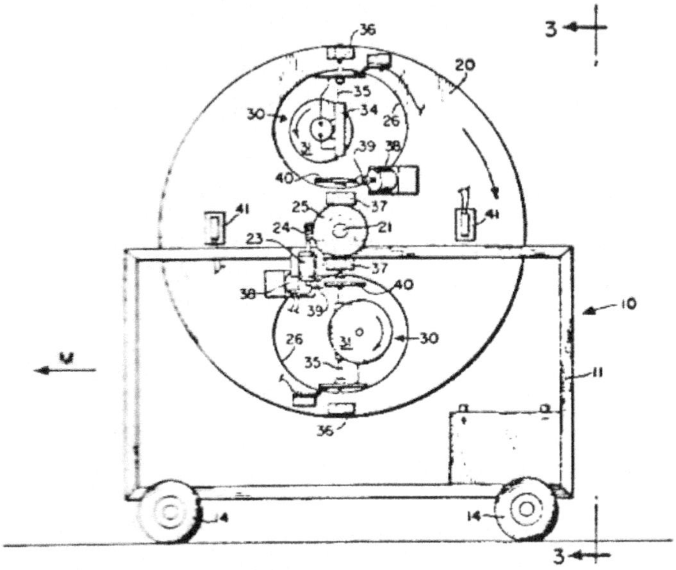

Fig. 5-21

Foster drive.

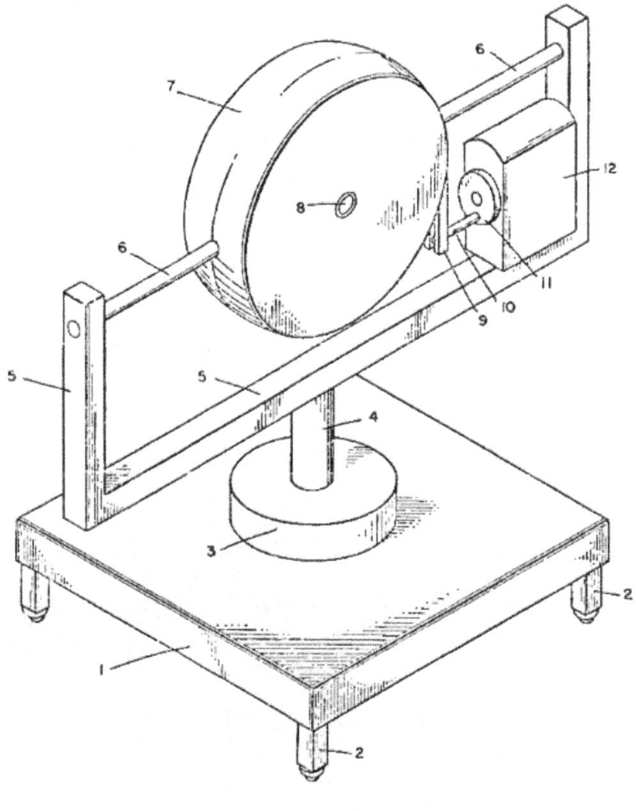

Fig. 5-22

Kellogg drive.

Many of the remaining drives each involve a unique principle. A complete analysis of them would be extremely difficult to present in simplified form. Some combine Coriolis, centrifugal, centripetal, and gyroscopic forces into super sophisticated machinery. Still others involve electromagnetic forces beyond the scope of this book to address. Table 5-1 lists selected patents.

These inventors are pioneers. Facing the established scientific community with ideas certain to incite controversy (or worse) takes courage and stubbornness. We honor these inventors.

Newton's laws of motion and the laws of conservation of momentum have prevailed against these machines—up until now.

TABLE 5-1
Selected Patents & Their Principle Propulsive Forces

INVENTOR	*MAIN PROPULSIVE FORCE*
Llamozas	impulse
Kellogg	gyroscopic
di Bella	Coriolis, centrifugal
Auweele	impulse
Foster	gyroscopic
Young	gyroscopic, Coriolis, centrifugal
Lehberger	centrifugal
Cook (1972)	Coriolis, centrifugal, centripetal
Cook (1980)	centrifugal

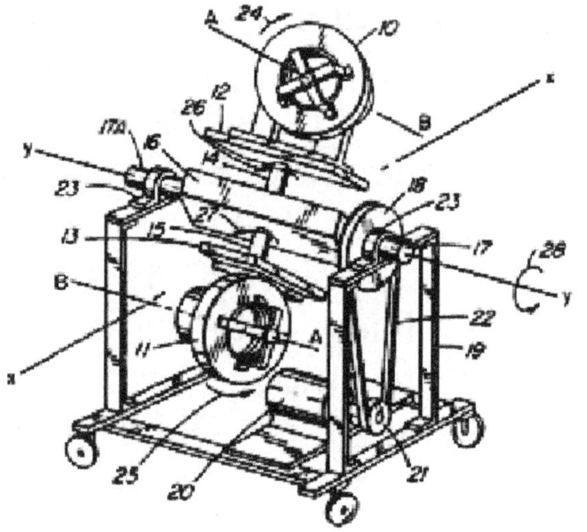

Fig. 5-24

Delroy drive.

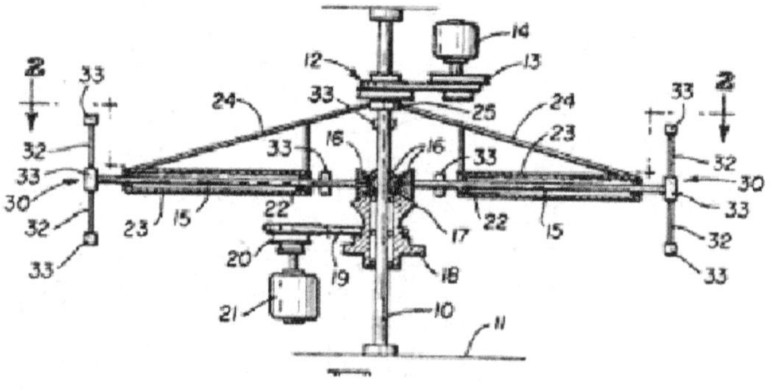

Fig. 5-23

Young drive.

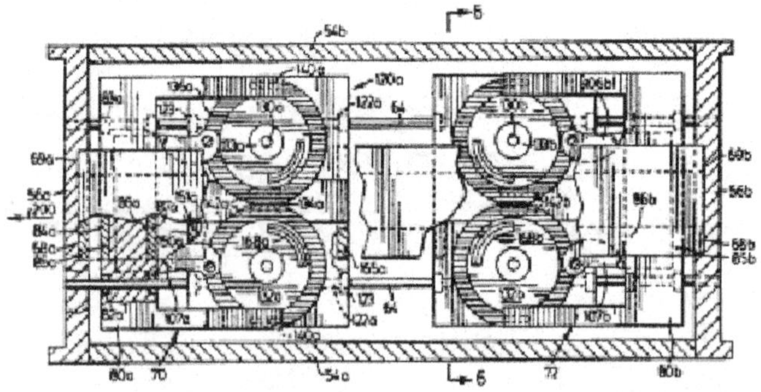

Fig. 5-25

Black drive.

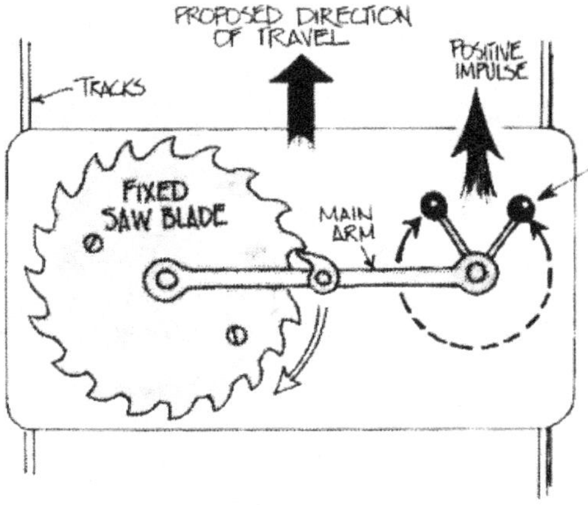

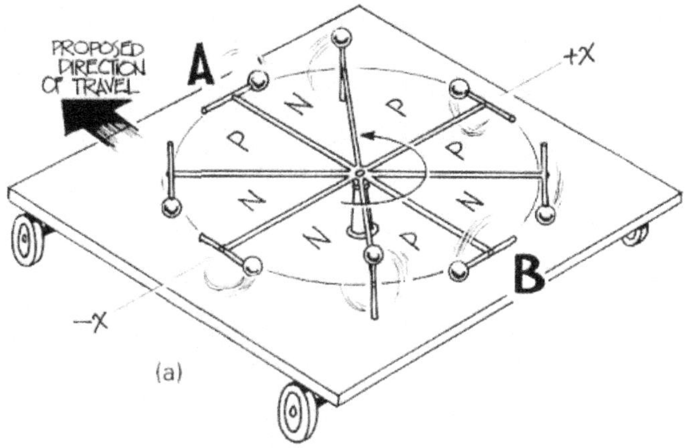

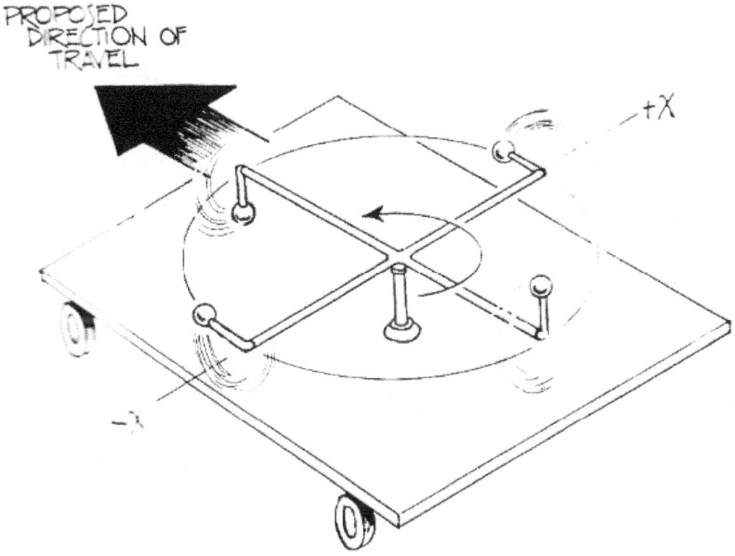

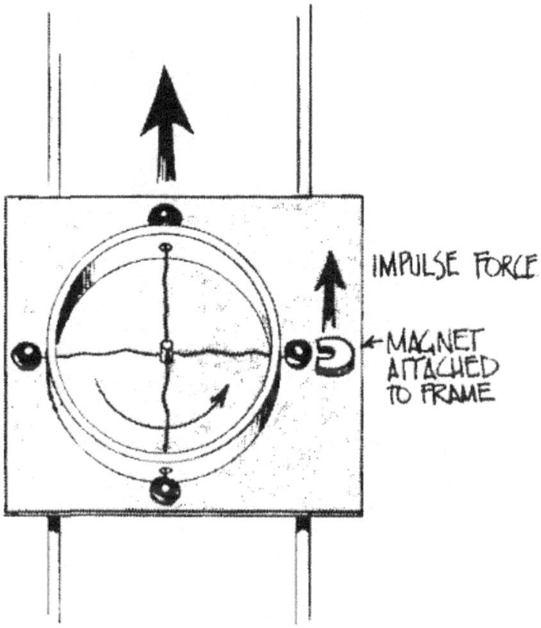

A few of our own unsuccessful reactionless drive experiments.

6

THE DISCOVERY

In 1968, Bob Cook first began research that will unlock some of the greatest mysteries of the universe.

What sparked this research was a book Bob had written about his personal spiritual experiences that he wanted published. Being unknown, however, no one took his book seriously and he was advised to gain public attention in order to generate a base of interest.

To do this he decided to attempt to invent something considered impossible, since nothing mechanical had ever permanently stumped him.

To this end, he originally set out to build a perpetual motion machine to generate electrical power. While working on this, he made an unintentional change to a rotor that made him realize the model was going to propel. Convinced that a propulsion system would be easier to promote than perpetual motion, he decided to resume his energy experiments later. He committed himself to proving and getting public acceptance of his new idea: The Cook Coriolis (CC) drive.

The concept became reality in the spring of 1969. Bob built CC-1, a small, single-rotor prototype powered by a 1/70th horsepower electric motor. As he predicted, the spinning rotor produced about 9 ounces of unbalanced force, which propelled the 10-pound wheeled cart on which it was mounted (Fig. 6-1).

A single-rotor model is limited to surface propulsion. A system capable of propelling in deep space would require at least two counter-rotating rotors to cancel gyration and forces lateral to the desired direction of travel (Fig. 6-2).

By that summer, Bob started constructing CC-2, a hand built four-rotor device that, when completed, would be capable of propelling in *any* environment.

After several months of working on the new machine Bob still had not made it propel. Even through his

The single-rotor CC-1 proved the concept.

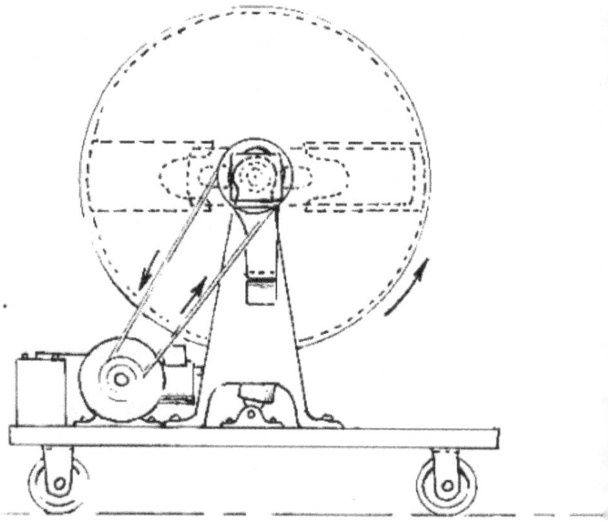

Fig. 6-1

Schematic drawing of this early CC concept.

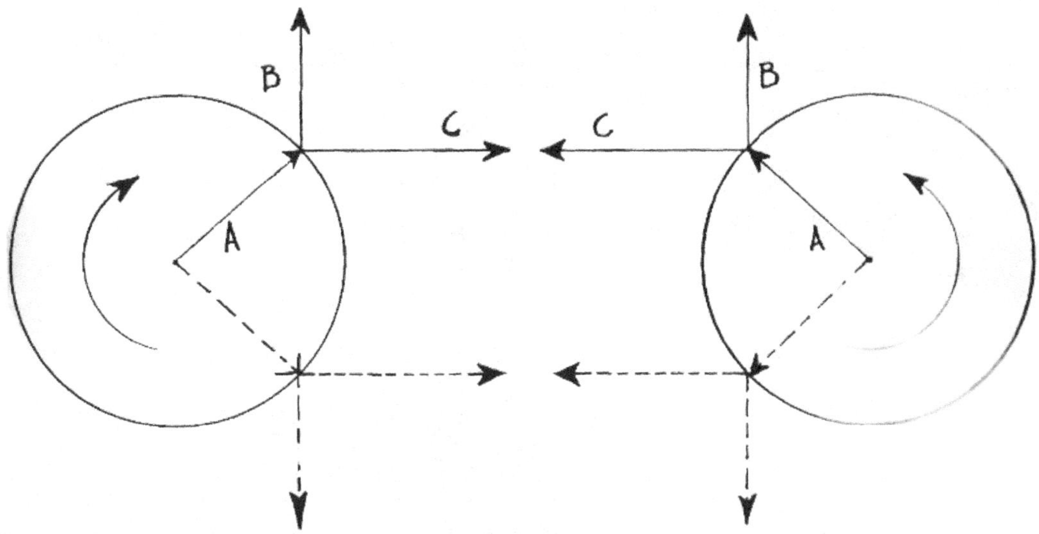

Fig. 6-2

The cancellation of horizontal forces produces a two-directional force

frustration, though, he would not quit until he found the way to make it work. In time, after much hard concentration and troubleshooting, he finally made the necessary adjustments and the machine, for the first time, began to propel. His excitement at seeing CC-2 move forward was overwhelming and that evening he made several phone calls to relatives to tell them of his tremendous success.

This second model was crude, so in the spring of 1970 Bob built CC-3, an improved two-rotor version. Aided by this working model, he filed for a United States patent using the twin counter-rotating rotor concept.

The handmade, four-rotor model CC-2.

CC-3 was built in Arizona in 1970.

The eight-rotor CC-4 was built in Texas in 1971.

All three of these models propelled in a series of surges. Some professors who witnessed his first public demonstrations held at the University of Arizona in Tucson claimed this was proof the device moved because of the "stick-slip effect," exchanging momentum with the floor through friction. Bob reasoned that by adding another set of rotors phased at proper intervals he could fill more of the dead part of the cycle with positive force and thereby produce a constant force. Such a machine would move smoothly and even accelerate.

So in early 1971, the eight-rotor CC-4 model was built. Unfortunately, eight rotors was not enough, and the machine again propelled in surges.

During the summer, unsatisfied with the complexity and lackluster performance of CC-4, Bob completely redesigned it and made CC-5, a better-running, four-rotor unit.

Through a friend at *Gazette Press* in Berkeley, California, Bob made contact with the engineering department of United Airlines. On September 10, 1971, Bob took his CC-5 model to United's main test center near San Francisco where a dynamics analysis was undertaken.

Bob was surprised by the interest his machine created during this initial visit. He found himself demonstrating his model to over twenty of United's personnel. Several of the engineers who witnessed the device propelling across the floor expressed enthusiasm and amazement at the implications.

CC-5 was studied by United Airlines.

The dynamics analysis took several months to prepare. The introduction stated, “In spite of being declared in violation of the laws of motion by the United States patent office, Cook’s crudely built rig moved spasmodically across the floor.”

The United report (Ex. 6-1) concluded that, although weak and inefficient, the machine did produce a net positive thrust!

Let’s at this point look at the invention studied in this report (Fig 6-3), which was patented in 1972.

Two counter-rotating rotors each consist of a carrier containing a lead bar (propellant mass) shorter than the carrier. The counter-rotating rotors are phased so that the forces lateral to the direction of motion are canceled. The carriers are mounted on shafts connected through bearings to the main frame of the vehicle, and are rotated at a constant angular velocity. The frame is mounted on small wheels. At the ends of the carriers are springs attached to the frame. These are used to accelerate the propellant mass radially.

Forty-five degrees from the desired direction of travel both lead bars are accelerated radially in the carrier and impact in the rearward direction (Fig 6-4). Then they are recycled to the 45° position, and the action repeats.the 45° position, and the action repeats.

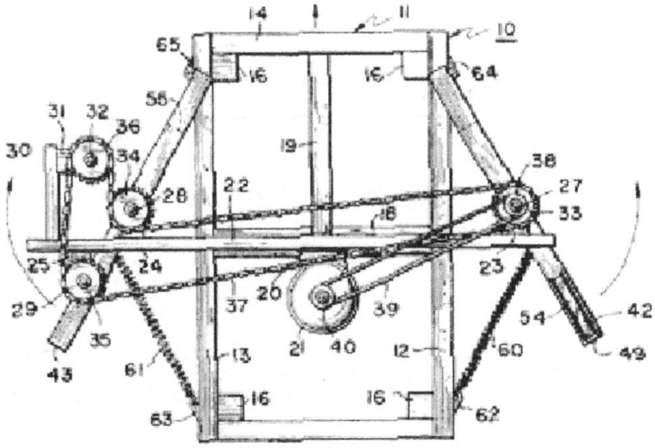

Fig. 6-3

Cook Coriolis drive.

Ex. 6-1: Excerpts from United Airlines' 26-page analysis of the Cook Coriolis drive.

United Air Lines

ROUTING							

TEST CENTER AND PROCESS ENGINEERING
SERVICE REQUEST
85/86/-45-08-00-01

NUMBER D-71-77
DATE 9/10/71

REQUEST SUBJECT DYNAMIC ANALYSIS
FILE 00-00-00

DESCRIPTION: (Reason, Priority, Statement, Sketch, Etc)

Provide a dynamic analysis of the propulsion device designed by Robert Cooke.

ORIGINATOR | PHONE NO. | DATE | APPROVED BY MANAGER | DATE

FOR TEST CENTER AND PROCESS ENGINEERING USE ONLY

REPLY ~~See attached~~

ENGINEER | 11/11/71 DATE | MANAGER | DATE

8

INTRODUCTION

On 9/10/71 Robert Cooke brought to UAL a device designed to convert centrifugal force into a linear thrust. In spite of being declared in violation of the laws of motion by the U.S. Patent Office, Cook's crudely-built rig moved spasmodically across the floor.

This report provides a dynamic analysis of Cook's mechanism. The cycle demonstrated by Cooke, as well as two other cycles which offer performance improvements, are examined...

Cooke's Propulsion Cycle

Cooke set up his working model so that the propellant mass followed the path shown in Figure 3. From point 1 to point 2 the propellant mass is pinned against the end of the tract by centrifugal force. The thrust seen in this segment is the component of centrifugal force in the direction of the cart motion. This thrust is

$$T_{1-2} = m_p R\omega^2 \cos \omega t \qquad (1)$$

where m_p is the propellant mass,

R is one half the sliding distance of the track,

ω is the angular velocity of the rotor, and

t is time.

Due to Cooke's positioning of the spring, the propellant mass spends more time behind the center of rotation of the track than forward of the center. Thus, the net thrust in segment 1-2 is negative.

When the propellant mass reaches point 2, the spring force overcomes the centrifugal force, and the mass accelerates down the tract to point 3. During this portion of the cycle the system acts as a mechanical analogue of a rocket. The propellant mass is accelerated in the aft direction by the spring force and the resultant reaction produces a forward thrust upon the cart. In addition to this reaction force there is Coriolis force which is the inertial effect occurring when a mass is constrained to move in a straight line across a rotating body. The total thrust in segment 1-2 is

$$T_{2-3_a} = 2KR\cos\lambda t\cos\omega t - \frac{4K\lambda\omega m_p}{2K - m_p(\lambda^2+\omega^2)}\sin\lambda t\sin\omega t$$

where K is the spring constant, and

$$\lambda = \left[-\frac{\omega^2}{2} + \frac{K}{2}\left(\frac{1}{m_p} + \frac{1}{m_o}\right) + \frac{1}{2}\sqrt{\omega^2\left[\omega^2 - 2K\left(\frac{1}{m_p} - \frac{1}{m_o}\right)\right] + K^2\left(\frac{1}{m_p} + \frac{1}{m_o}\right)^2}\ \right]^{1/2}$$

m_o is the mass of the cart.

At point 3 the propellant mass strikes the end of the cart producing a negative impulsive force.

$$T_{3-1} = -\frac{2}{\Delta t}\left(F_0 + \frac{m_p^2 R\omega^2}{m_0 + m_p}\right)\sqrt{\frac{R}{KR/m_0 + F_0/m_p}}\ \cos\omega t \qquad (3)$$

where Δt is the time required to stop the propellant mass, and

$$F_0 = KR - m_p R\omega^2$$

During this segment of the cycle the propellant is stopped at the expense of the forward momentum of the cart.

The resultant thrust on the cart for the entire cycle is shown in Figure 3.

A Modification of Cooke's Cycle

A significant improvement in performance can be achieved by using viscous damping to arrest the propellant mass. Not only can the large negative impulse be avoided, but by delaying the travel of the mass to the end of the track, the negative centrifugal force componant can be reduced...

...Cooke's cycle could also be improved by the use of a constant force rather than the variable spring force to accelerate the propellant mass. This would increase the thrust during the ejection stroke by allowing the use of greater force and improving the timing of the stroke...

Mechanically, it's rather simple, but the full force picture is extraordinarily complex. Moving the bars radially causes two Coriolis impulses to occur simultaneously in each one. A positive Coriolis impulse is produced by the portion moving toward center, and a negative Coriolis occurs in the portion moving away from center. Because of their quadrant, the force vectors are additive in the desired direction of travel. This principle demands simultaneous Coriolis forces.

A good way to visualize this is in the form of a cone—shown by the concentration of dots in Figure 6-5. When the greater mass moves toward the center it has to give up energy (which produces a positive Coriolis force) while the mass moving away from center increases its energy content and produces negative Coriolis force. The Coriolis forces combine in the positive direction with the increased centrifugal force imbalance created by the removal of the centripetal force in the short end of the mass.

To complete the picture there are other forces (such as gyroscopic forces) to consider. It looks so innocent but yet is super complex. No one has ever fully analyzed the complete principle.

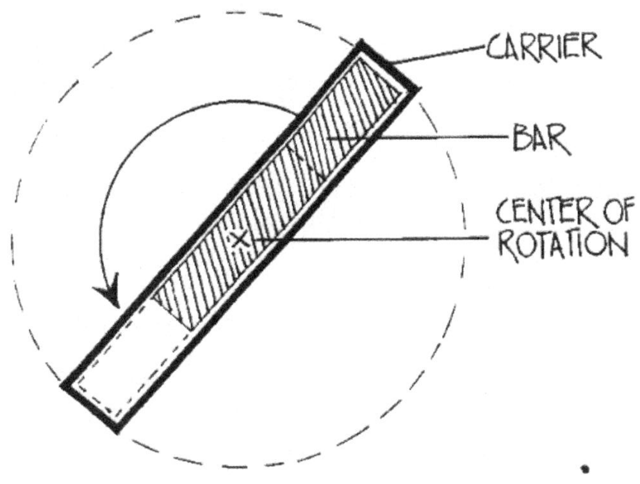

Fig. 6-4

Lead bar properly positioned in its carrier.

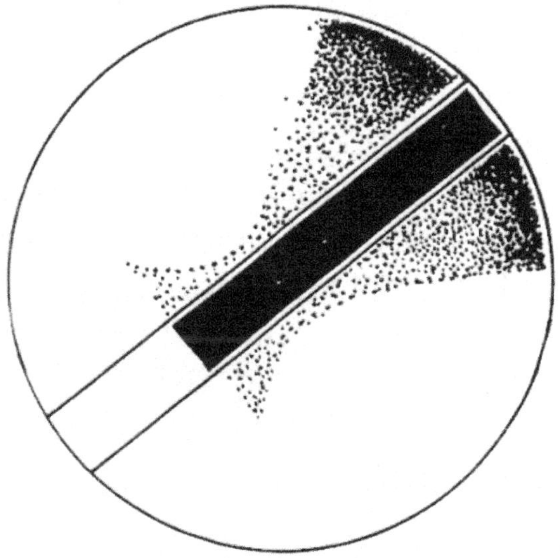

Fig. 6-5

The inertial force cone.

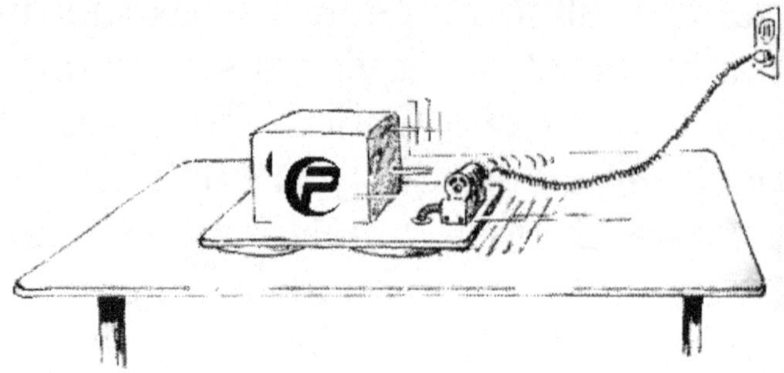

Fig. 6-7

CC drive on an air cushion propelling on a flat laboratory table.

Many science experts theorized that Bob's models took advantage of static friction to impulse along. With friction removed, they reasoned, the devices should simply oscillate back and forth.

Bob conducted several experiments suggested. Perhaps the wheels were not properly lubricated? A few squirts of oil and the machine moved beautifully. What would happen if the device floated in water, with no floor with which to exchange momentum? Bob tested it on a small raft and away it impulsed. Since there is still some friction in water, what if this were eliminated by putting it on finely-honed hockey blades and running it in an ice skating rink? It surged across the smooth ice, dragging its 200 pound inventor with it! Maybe the device reacted against the surface upon which it rested in order to propel? Bob bought air cushions, mounted the model on it, and it worked perfectly (Fig. 6-6).

To his expectation and many others' disbelief the CC drive worked *better* without friction.

To his expectation and many others' disbelief the CC drive worked better *without friction.*

In 1972 further tests were conducted at United Airlines. This time, the machine was tested for two weeks with an accelerometer attached to a polygraph recorder and oscilloscope (Ex. 6-2). The results showed a net force: 6¼ g positive to 3 g negative, with the positive area on the graph greater than the negative (Ex. 6-3).

United was willing to back up the analysis and its results to anyone who asked, but they would not commit to anything more. They were not in the business of promoting inventions.

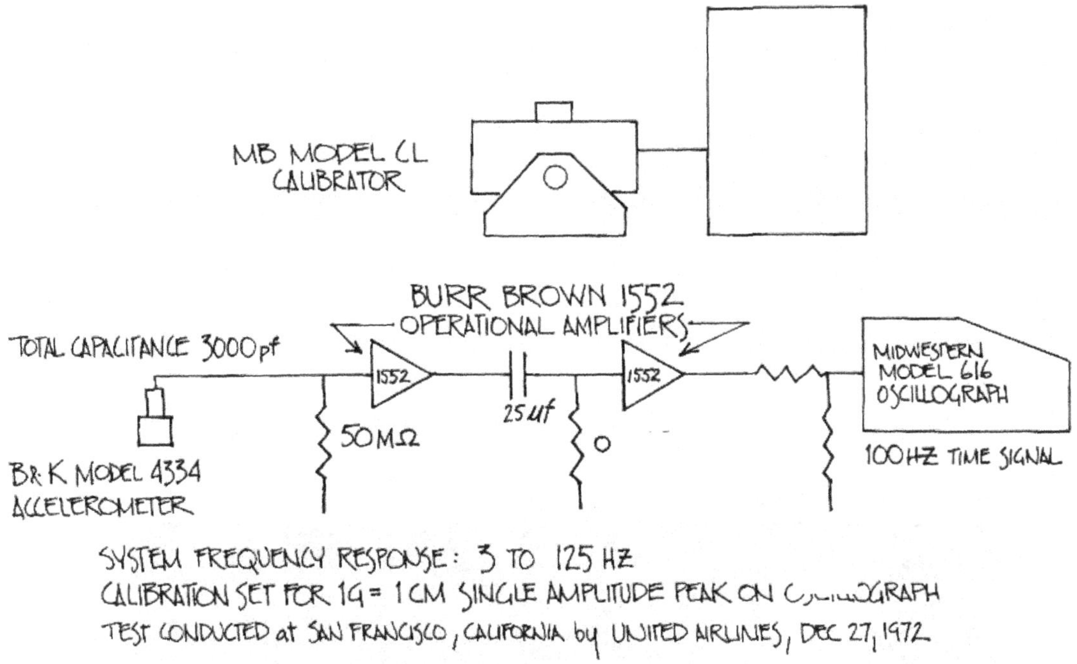

Ex. 6-2

Schematic of United Airlines'accelerometer test.

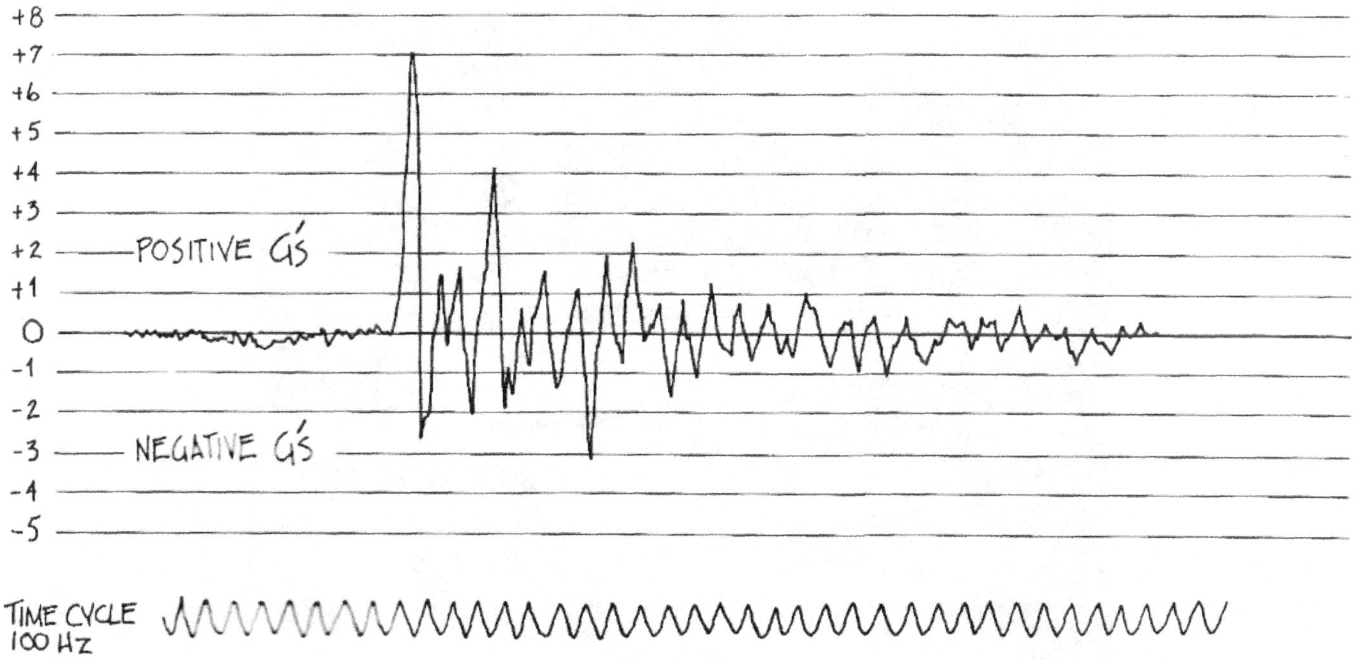

Ex. 6-3

Accelerometer printout of Cook's cycle showing positive conclusions.

Although these models had so far passed all tests and appeared to dispute the laws of physics, Cook was not satisfied with the intermittent force generated by the CC drive. He wanted a model that could produce a constant, unidirectional thrust. This would be much more convincing to the scientific establishment that so far had greeted most of his demonstrations with skepticism.

Bob estimated the CC drive had approximately a 1% propulsion efficiency. He took pride in the fact that the machine had made a tremendous discovery in physics, but he was not proud of this inefficiency.

Bob became determined to find a more efficient way to propel internally. In 1974, he abandoned further work on the CC drive and went back to the drawing board.

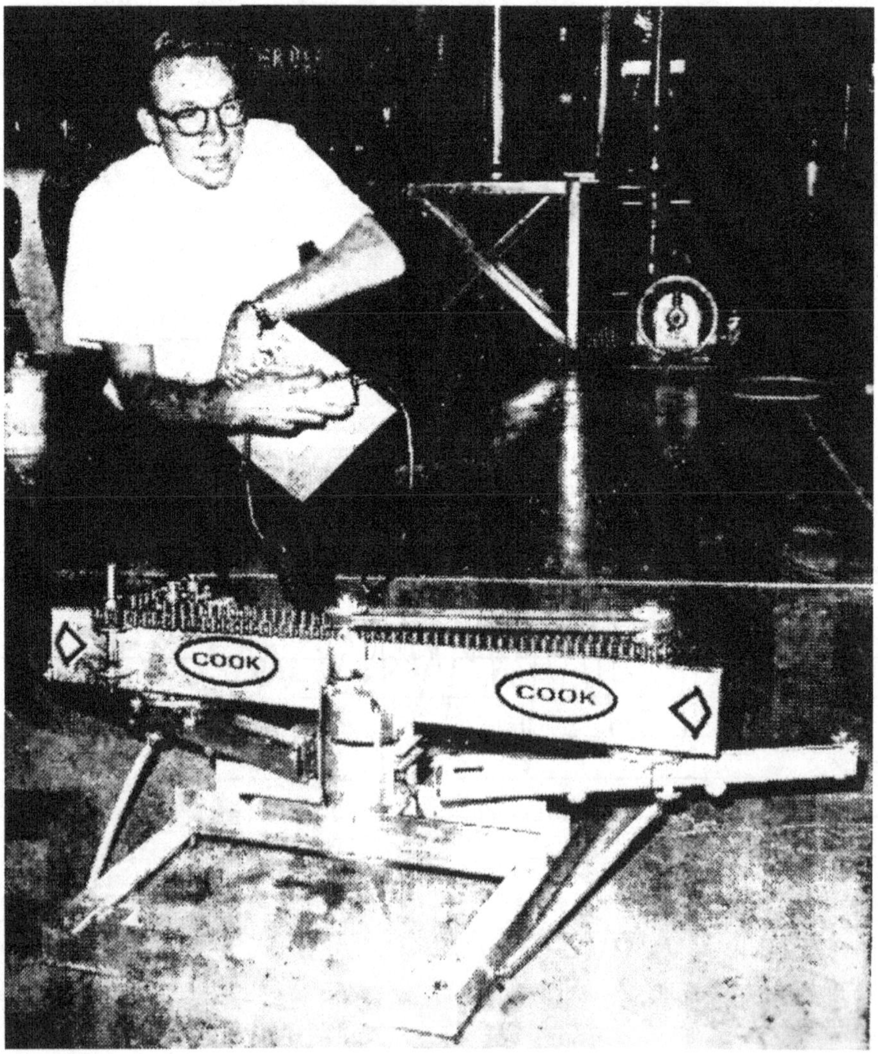

Bob demonstrating CC-3 at the University of Arizona at Tuscon.

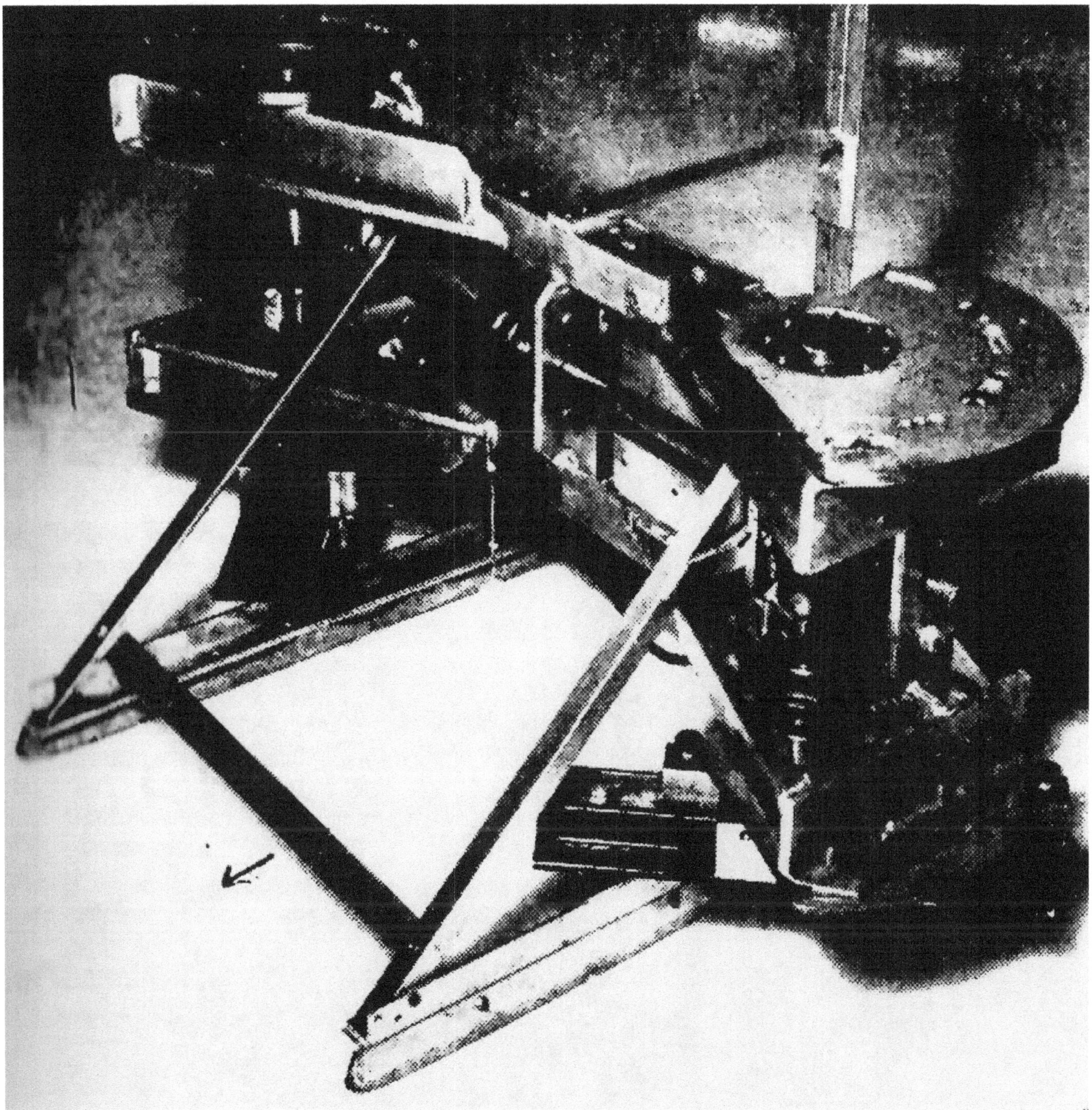

The professionally built CC-6 dragged Bob across an ice skating rink.

7

THE BREAKTHROUGH

In early 1974, Bob Cook began six months of intense experiments to find a more efficient way to propel. A basic approach to this soon developed: First identify any roadblocks, then circumvent these roadblocks. Before finding the right solution, he discovered many ways to not propel efficiently.

All known forces were explored for propulsion, including rotational forces, magnetism, and mass acceleration. Over a hundred possible combinations were considered, with models built to test some of the more promising ideas.

An elegant solution finally synthesized. By splitting the propellant mass for part of the spin, unbalanced centrifugal force could be harnessed for linear motion.

In December 1974 Bob returned to his patent attorneys to disclose the new idea. They were astonished and promised to give it top priority.

In the meantime, Bob decided to make a working model to test out his new theory, on which the *Vallejo Times-Herald* ran a story (Ex. 8-1). When attempting to hand-build a CIP engine like he had done with the CC drive, he soon learned that the exchange mechanism required to split and recombine the propellant mass required more precision than his own tools could provide.

Bob making adjustments on the first hand-built CIP prototype in early 1975

A second model was built by a friend with a small garage machine shop. However, this prototype also proved unworkable since the wrong ratios and proportions had been used. Even when these deficiencies were corrected, the model could not propel, yet it was not a total failure. Tests did confirm that is was possible for the exchange mechanism to smoothly transfer and reverse the direction of the mass between the arms in both directions.

These first models were limited by a small budget. Greater funding came in 1976 and a more precise model was made. Built with the correct dimensions, the new model demonstrated that the exchange mechanism would positively work at higher speeds and with much larger masses. It also confirmed that the nucleus mass, a critical aspect to the success of the CIP principle, was going to do its job. However, it proved difficult to keep the masses exchanging between the arms consistently enough to produce propulsion.

From early 1976 to the middle of 1978 progress was delayed for lack of funds. In July 1978, though, after another *Times-Herald* article was written, greater funding was secured and construction of the fourth CIP prototype started.

Inventor Builds The CIP Engine

By CAROL CROSS
Times-Herald Staff Writer

A new device which its inventor says could revolutionize propulsion systems on land, at sea and in outer space currently is undergoing tests and further refinement in the home of Robert L. (Bob) Cook of 605 Wilson Ave., Vallejo.

The Texas-born 41-year-old Cook arrived here two months ago from Pittsburg, and although his vocation is that of a printing pressman, he has been working full-time for the past three years and part-time for four years before that, on his rather odd-looking contraption.

The inventor, a bachelor whose formal education ended with his graduation from Mt. Diablo High School in Concord, has enough faith in the potential of his engine that he is having a new model built in Berkeley. He estimates his investment in the engine over the years to have totaled some $60,000.

Cook apparently took issue with Isaac Newton's third law of motion when he started out to perfect what he calls "the CIP engine."

"This highly controversial system, once considered totally unworkable by most scientists, is now beginning to get more attention from them and they are not so sure that it can't work," he said.

NEWTON'S LAW

"This controversy is caused by the fact that Newton's laws of motion imply that in order to move an object (or mass), you must either push it or pull it with an outside (external) force.

Robert L. Cook, of Vallejo, is an inventor working on a brand-new propulsion system that can be used on land, at sea and in outer space. Here he is seen with a mock-up of a portion of his device. He said it would take eight of these gadgets to complete his entire system.—Times-Herald Photo.

"The CIP engine is an internal force system which converts centrifugal force into a linear thrust. Experiments conducted with an early concept model, using highly sophisticated equipment at a modern laboratory, proved that a very weak internal force was being produced.

"Later experiments have led to a tremendous break-through and now show great possibilities."

As a result Cook sees "a ray of hope for the future at this time in history when so many people are trying to solve our energy and pollution problems without much success."

The inventor said the U.S. Patent Office granted a patent on the original concept in 1972 and that a new patent is pending on the "break-through" which he believes "could very well prove the CIP engine to be the world's most efficient system for propelling anything from submarines to spacecraft."

He adds that ce[illegible] recent experiments [illegible]ate such high efficiency can be achieved and that if so, the following things can happen:

1. "The cost of building and operating cars will go way down. Cars no longer will need transmissions, drive shafts and differential gears. The CIP engine car will be moved by an internal force similar to artificial gravity.

"It will use 35 to 50 per cent less fuel because of less moving parts and less friction."

2. "Surface vessels as well as submarines will benefit from this system. Submarines are plagued by the change of viscosity of water at great depths — which causes great resistance to the turning of their propellors.

"With this system, which doesn's require propellors to react on water in order to move, this problem will be eliminated."

3. "Space work is where the system could provide some science-fiction-like performances. Up until now the fastest man-made object, Pioneer 10, has reached a speed of 100,000 miles per hour.

"This is great speed, but compared to the speed po[illegible]l of an internal force [illegible]n it is nothing. Engi[illegible] who have studied the CIP engine believe that in space it could reach or exceed the speed of light (186,000 miles per second)!

HIGH SPEED

"One engineer calculated it would take the system two months to reach that speed (in space) if a constant 32.2 feet per second of acceleration was maintained." Some believe this is the type of engine used by certain UFOs, since they are not rocket-powered and the high-pitched sound they make undoubtedly is that of rotors turning at high RPM — just like a CIP engine!"

Cook describes his system as "very simple" and says this is how it works:

"It will consist (in most cases) of eight units synchronized and phased to produce a constant force. Each unit will consist of two counter-rotating arms, with one arm having a spinning rotor. This rotor will have a weight (mass) that can split in two.

"An effective [illegible]ng force will result (w[illegible] [illegible]e mass is together) for [illegible] degrees of arm travel and then the mass will split. Half of the mass will go clockwise and the other half counter-clockwise. This will neutralize the effective working force for 180 degrees of arm travel.

"The two masses are then rejoined on the spinning rotor and the cycle is repeated. The system can be powered by gasoline, electricity or just about anything that can make the arms and rotors spin.

"It will have almost unlimited force potential. To give an idea of this p[illegible]al — if you spin a on[illegible]d weight in a two-foot [illegible]eter circle at 7,000 RPM it will produce 14,000 pounds of force and at 40,000 RPM about 80,000 pounds."

"If you convert that into linear thrust, you can propel almost anything."

Cook is being assisted in his work on the engine by Joel Dickinson, 27, of Concord.

Ex. 8-1

The first published *Vallejo Times-Herald* story on the CIP.

Due to the illness of the original machinist hired to make the parts, it was not until May 1979 that Bob received all the completed major components. By summer, the single-rotor CIP-1 model was assembled.

Over the next few months the machine was fine-tuned to the point where a consistent exchange was possible at high speed. Late in the afternoon of October 15, 1979, Bob and an assistant made some final adjustments and gave CIP-1 its first true test.

The gears began spinning, the mass began exchanging, and the nucleus began to perfectly control the oscillations. Finally, the machine reached the necessary rotor speed required to overcome friction of the wheels—and the machine began to surge forward.

More adjustments were made to improve CIP-1 so it could complete a special experiment. On June 9, 1980, the model was mounted on a boat. Many experts predicted that on water this new mechanism would only produce bounded motion. This time, the machine not only propelled the length of the pool in continuous motion, but for the very first time actually accelerated.

History had been made.

On June 9, 1980 CIP-1 accelerated this boat across a pool.

In December 1980 a United States patent was issued for the CIP principle. Over the next several years the invention was also patented in Canada, the United Kingdom, France, Germany, and Japan.

After this proof, Bob focused on funding an improved prototype that could give an impressive demonstration to the public. In 1981 new investors were found who set up a corporation and raised money to build CIP-2, an advanced twelve-rotor model. Designed by hired engineers who did not heed Bob's advice, it could only exchange at slow RPM and generated a weak force. Nonetheless, it propelled on air bearings over a perfectly flat micro top table—a feat only matched by other Cook-built drives.

In 1983 this promising development suddenly turned into a nightmare. One of the main promoters of the CIP engine embezzled all of the company funds and fled to Canada, leaving Bob embroiled in a lawsuit with several investors. It was eventually resolved in Bob's favor by 1988.

In spite of this catastrophe, Bob steadily continued to refine the CIP engine. CIP-3, a highly sophisticated six-rotor prototype completed in 1987, was strong enough to propel itself while hanging from an upward inclined track.

CIP-2 produced a force so weak it could *only* propel on air bearings.

CIP-3 propelled up this inclined track.

By the mid-1990's he found new investors and began designing a new prototype powerful enough to demonstrate the practicality of using a reactionless drive to propel. Although it was never fully completed, the large two-rotor CIP-4 was tested with accelerometers, which showed a net positive force.

A novel experiment was also attempted with this machine. At the suggestion of another scientist, Bob built a Cavendish torsion pendulum large enough to mount CIP-4. Such a horizontal pendulum is sensitive enough to detect the faint gravitational attraction between objects of various sizes. Such a test would easily determine if any unbalanced force was being produced by the system.

As had happened before, the CIP engine proved itself. With each cycle the model deflected around the center of the pendulum several inches. This far exceeded the fractions of an inch expected by the expert.

CIP-4 is the largest successful reactionless drive built to date.

CIP-4 during accelerometer testing in 1999.

8

THE PRINCIPLE

Until now the principle of the conservation of angular momentum has stood the test of time, judging by the many failures experienced by inventors attempting to create reactionless drives. In most cases, in trying to destroy angular momentum in order to propel, these inventors have succeeded in creating an equivalent negative linear thrust that destroyed any positive motion.

Here is the secret to how the Cook Inertial Propulsion engine circumvents this problem.

HOW CIP WORKS

The propulsion unit itself must be considered a closed system. The motor, being a separate part of the system, can be considered an open system. Bear in mind that the propulsion system requires a prime mover, such as electric motors, gasoline or steam engines, and so forth.

Each unit has two counter-rotating upper and lower arms maintaining a constant angular velocity. Attached to the end of the upper arm A is a small rotor C, which is geared to spin at a constant velocity. It spins in a plane perpendicular to both arm A and arm B. The small rotor C has a permanently fixed mass (FM) attached at one end, an identical exchangeable mass (EM) attached at the opposite end (both together called the propellant mass), and a nucleus mass (N) attached to the center of the rotor. At one end of arm A (opposite of rotor C), a counterweight exactly balances the nucleus mass. (Fig. 7-1)

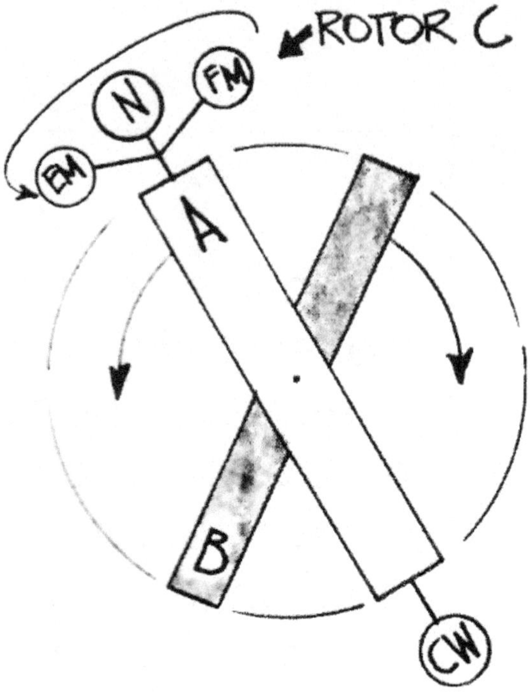

Fig. 7-1

Counter-rotating arms with rotor and counterweight.

The exchangeable mass transfers from rotor C to lower arm B, splitting the propellant mass. After 180°, it is transferred back to rotor C. This cycle (Figs. 7-2a through 7-2d) is continuously repeated to create an imbalance in the system. The object of this system is to synchronize rotor C and arm B so that the exchangeable propellant mass can be transferred without creating an impact.

There are only so many forces in nature that can be produced when something spins. The three main forces are centrifugal force, Coriolis force, and gyroscopic precession. The gyroscopic precession that develops in this case is vertical. All Coriolis forces generated are horizontal. These undesirable effects are cancelled with a complete, four-rotor system (Fig. 7-3).

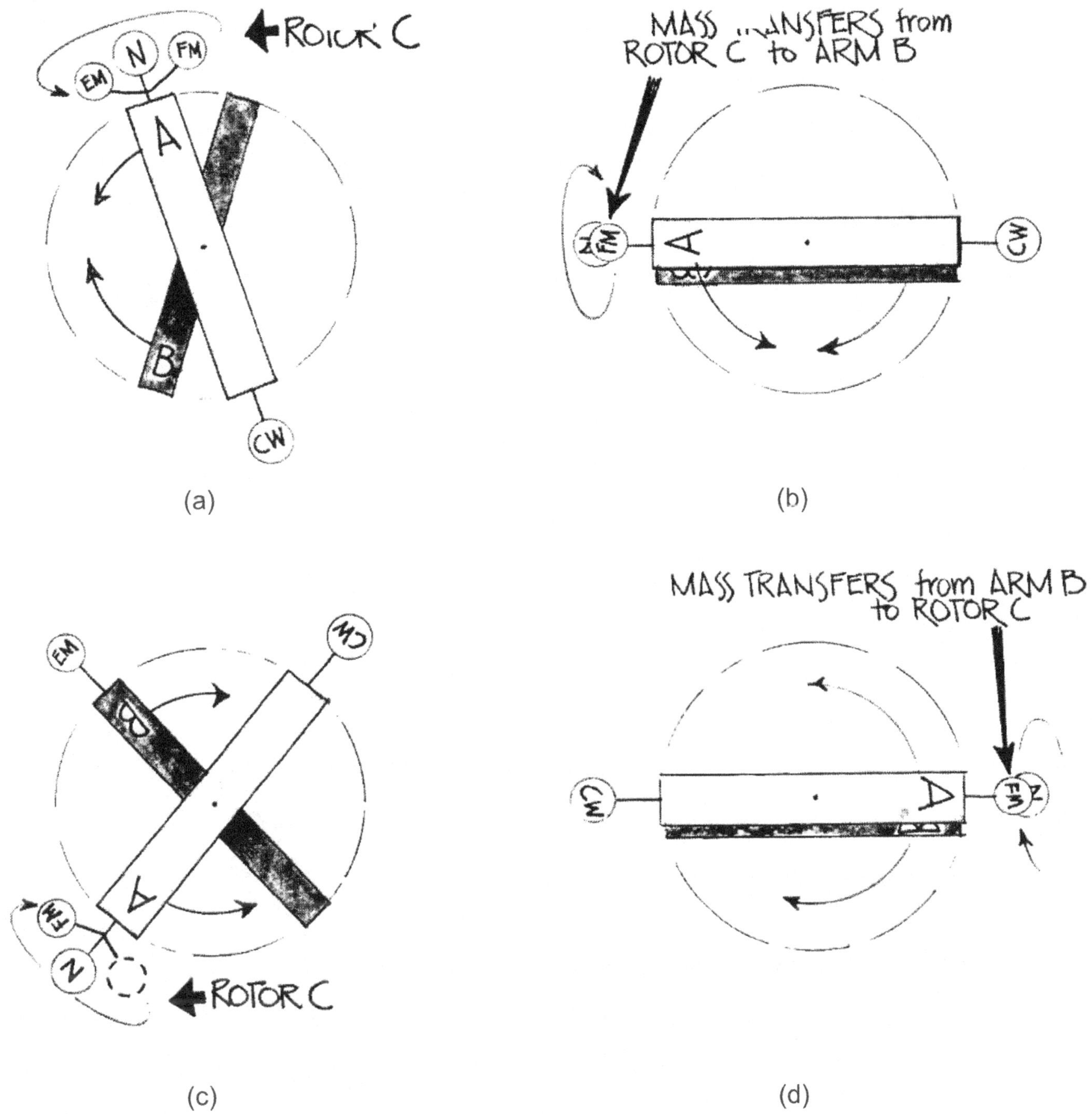

Fig. 7-2

Cook Inertial Propulsion drive.

(a) Positive force cycle; (b) mass splits; (c) neutral cycle; (d) mass recombines.

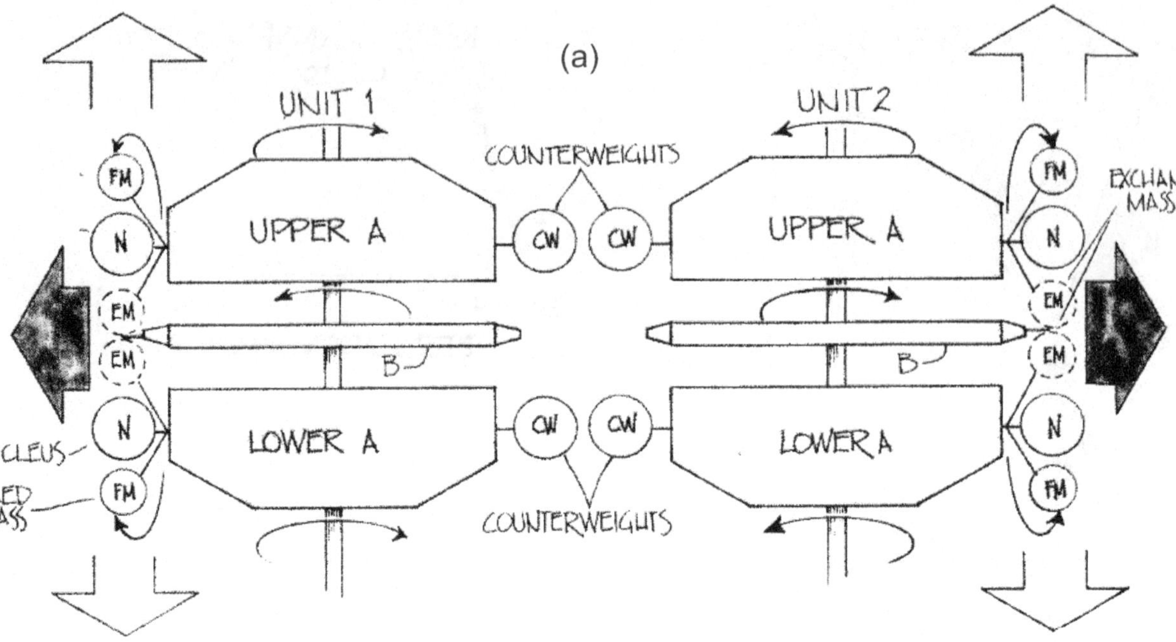

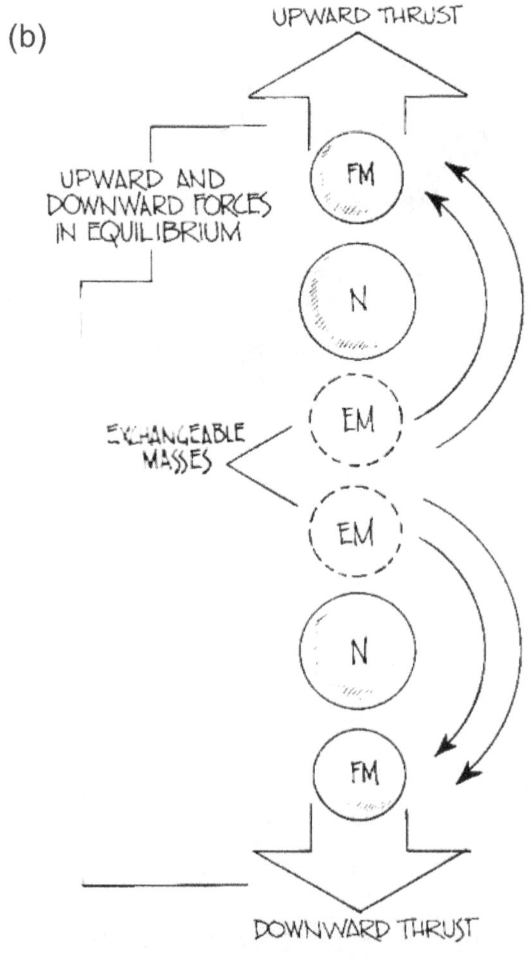

Fig. 7-3

(a) Unwanted vertical and lateral forces canceled by four counter-rotating rotors.

(b) Two counter-rotating rotors cancel unwanted vertical forces.

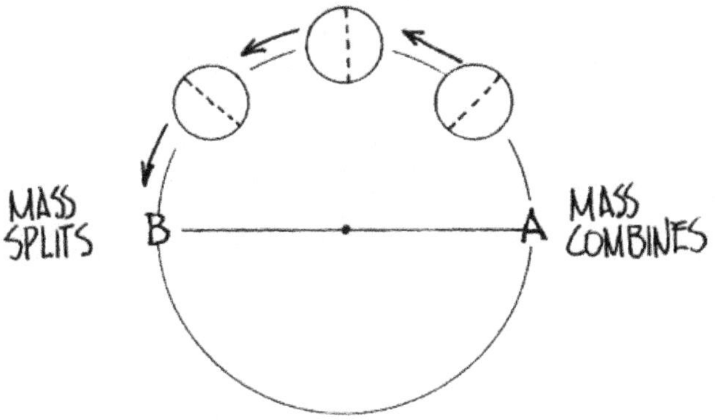

Fig. 7-4

Masses combine for 180° to produce positive cycle.

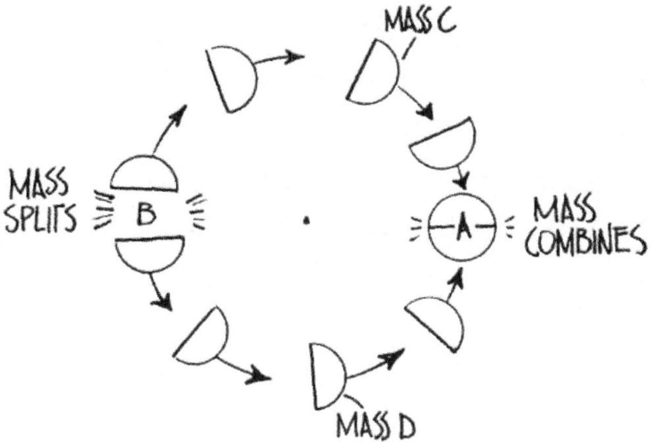

Fig. 7-5

Masses split for 180° to produce neutral cycle.

The only force remaining to consider, then, is centrifugal force.

For 180° between A and B we have a complete propellant mass producing positive centrifugal force (Fig. 7-4).

For 180° between B and A we have split the propellant mass. The centrifugal force generated by mass C equals the centrifugal force of mass D. We have 180° of forces in equilibrium (Fig 7-5).

Now let us consider the independent path of travel for each mass during a complete revolution of the main arms. The centrifugal force generated by the permanently fixed mass on the rotor balances (or cancels) over 360°. The exchangeable mass, however, is forced to remain on the positive 180° of arm movement producing positive centrifugal force except at two points, where its direction is reversed.

A vector analysis of the exchangeable mass does show a negative impulse force produced when the direction of mass is reversed. This negative impulse has a much shorter time to effect negative motion than the positive force produced by the combined propellant mass. Nonetheless, this negative impulse could completely cancel the longer lasting positive force unless rectified in some way (Fig. 7-6).

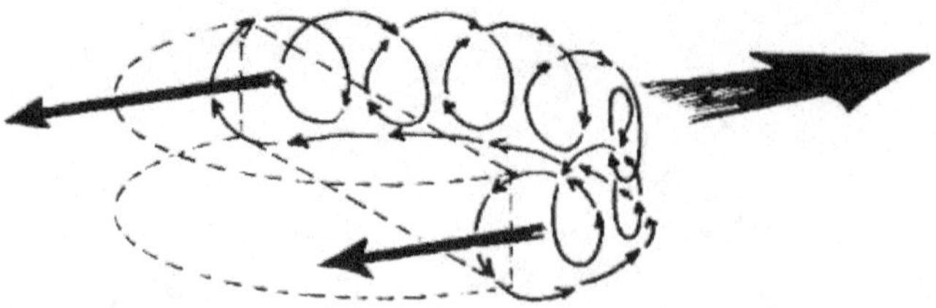

Fig. 7-6

The two intense impulse forces produced by reversal of the mass cancel the longer lasting, less intense positive centrifugal force.

To do this, flexibility of the system is required. The negative impulse is controlled by the nucleus mass located at the center of the rotor. The nucleus mass absorbs the negative force and energy instead of allowing these factors to produce a negative effect on the overall system. When the small rotor is out of balance, this flexibility also compensates for the undesirable effects by the unbalanced centrifugal force on the rotor by allowing oscillation.

The oscillator feature allows the necessary flexibility needed to complement the action of the nucleus mass, which is to temporarily absorb and store this energy. This allows the unbalanced centrifugal force to advance and retard the rotor (along with its nucleus mass) up to several inches by trapping the rotor's drive shaft in a motion-limiting slot (Fig. 7-7).

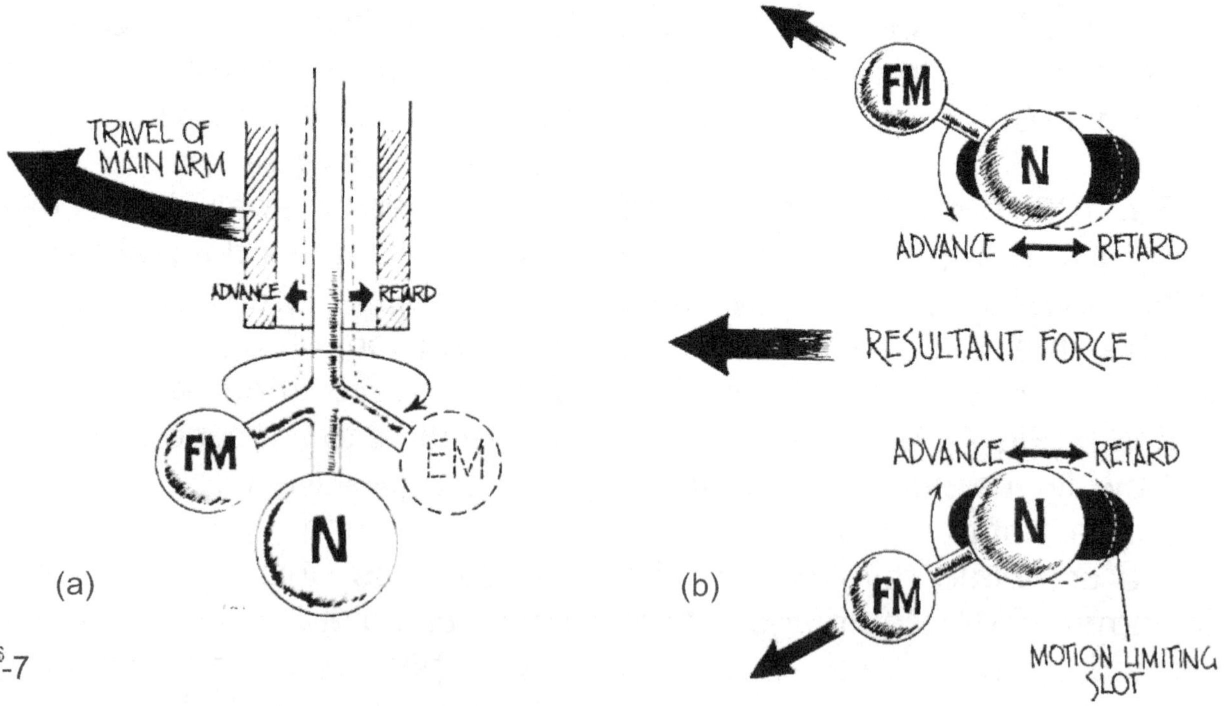

Fig. 7-7

(a) The unbalanced rotor oscillating; (b) the motion-limiting slot canceling the effects of resultant force.

Without the nucleus mass, the energy in the fixed mass would produce a negative force that would translate to the center of the system, destroying the positive motion imparted by the combined propellant mass cycle.

THE LOOPHOLE

Spinning masses are subject to the law of conservation of angular momentum. When no external torque acts on an object or a closed system of objects, no change of angular momentum can occur. When this was first formulated centuries ago, the assumption was that the mass would remain constant throughout the full 360° of spin.

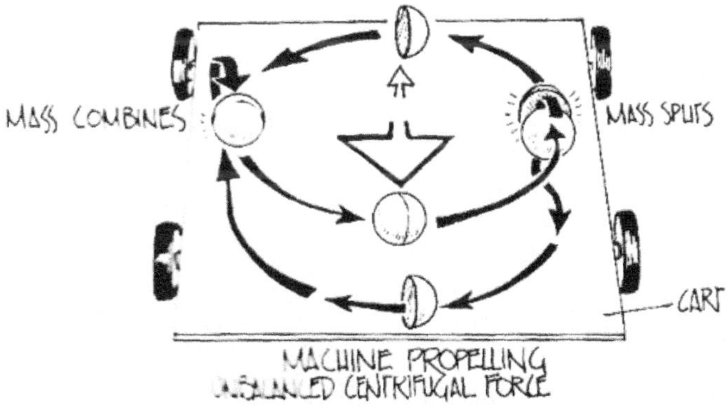

Fig. 7-8

The CIP engine splits its propellant mass for 180° of its spin.

As you have seen, the CIP engine mechanically splits the mass after 180° of spin (Fig. 7-8). One half of the equally split mass reverses direction. Because the evenly divided masses move at the same angular velocity, they create forces in equilibrium for only 180° of the complete cycle. These split masses are then recombined, becoming one mass which creates an unbalanced centrifugal force for 180° which can be used to propel. Through precise synchronization and flexibility, the splitting and recombining of the masses does not create negative impulses that could cancel the positive force rectified for motion.

That the law of conservation of angular momentum can be circumvented is a discovery with profound implications. When this phenomenon is more fully analyzed, the findings will expand our understanding of physics, mechanics, conservation of energy, and thermodynamics.

9

EPILOGUE

In the midst of developing the CIP engine, Bob Cook married Scherl Carr in June 1976. Over the next several years their family grew to include four boys: Rob, Jr., Victor, Benjamin, and Joseph.

These children grew up in a unique environment, sharing in the excitement and frustrations their father experienced while he worked to prove his reactionless drive to the world. Confident that someday success would come, they helped as much as they could to achieve this goal.

Rob, Jr. paid especially close attention to all of this. In his youth he assisted his father in the shop, accompanied him on business trips, and even joined him on television shows. He met all manner of people, saw how deals were made and broken, and learned firsthand how tough taking on the establishment really is.

In 1989, father and son built CID-0, a working CC drive that propelled Rob and his brothers as they stood on it. Those who witnessed it were impressed. This experiment also won honorable mention at his middle school science fair.

Although he shared with his father a knack for designing and building things, Rob spent his teenage and early adulthood years focused on writing. In the meantime, he worked odd jobs to support himself and his own growing family. Rob figured that once the CIP project became successful, he would be most useful as the creator and editor of communications for the family business.

But in 2004, Rob grew concerned that his father's failing health might bring a premature end to the work. Gaining his father's blessing and assistance, Rob set out to build a new CIP prototype.

The Cook family in 1980. From left to right: Victor, Scherl, Bob, and Rob, Jr.

CID-0 propelled Rob and his brothers across the ground.

Rob experimented with new methods of harnessing the CIP principle. Rob spent many months considering ways to implement certain improvements into a working device. One night in October 2004, after discussing it with his father, the solution became clear. When Rob explained his idea to his father, Bob was impressed and agreed that this new CIP cycle would work. To differentiate it from the original CIP system, it is referred as the Cook Force (cForce) drive.

Rob and some of his friends pooled their resources and incorporated cForce, Inc., in June 2005. Over the next two years, Rob designed and built cForce-1, a four- rotor prototype (that could be upgraded to eight rotors) and designed to generate over two hundred pounds of propulsive force. By March 2007 the first successful exchanges were made, but funds ran out before the final modifications to make the model propel were completed. The cForce project ground to a halt.

In the midst of this, the elder Cook's health began to rapidly deteriorate. Motivated by his deep love for humanity and urgent concern for the environment, Bob had never wavered in his efforts to introduce his inventions to the world. He did this in spite of limited funding, the skepticism of mainstream science, and the outright dishonesty of some people who could have assisted with this momentous project.

Rob with cForce-1.

This hard-fought struggle eventually caught up with him. On the afternoon of September 8, 2008, Bob Cook passed away in Bakersfield, California. He was 74.

Bob's passing was not only a terrible personal loss, it was an incredible loss to inertial propulsion research. Rob felt that to continue the work, he needed to gain the deep understanding his father had of the inertial forces and how to rectify them.

Not letting a lack of funds stop him, Rob returned to the old, simple CC drive for his initial experiments. He built the very crude CID-1 out of wood and spare parts over a weekend in June 2009. It only worked a few times before the rotors broke down, but it did propel and these tests were recorded on video which Rob studied.

Combining this and other hands-on experiments with his lifelong experience with reactionless propulsion, Rob began to develop a deep, intuitive feel for these forces.

Rob built CID-1 to study how to harness inertial forces to propel.

A totally new concept for reactionless propulsion soon formed in his mind. Rob spent the rest of the year expanding on this new concept, which he called the Cook Inertial Drive (CID).

(This acronym originally meant "Coriolis Impulse Drive" in CID-0 and CID-1. These devices were actually CC drives and did not utilize this new propulsion principle.)

He designed a new model that could be built from some spare parts no longer needed for cForce-1. By March 2010, CID-2 was basically complete, but several weeks of modifications and initially unsuccessful experiments followed. Then in the early morning hours of May 2, 2010, the right adjustments were found.

This time, when CID-2 was spun up, it immediately scooted forward. Rob watched in elation until the little model reached the end of its power cord and actually unplugged itself! Another method of inertial propulsion proved successful.

Rob tested CID-2 extensively, recording his experiments and studying the results. Based on this data, he designed the improved CID-3, which was built by the spring of 2011. It first propelled in April and by that summer was moving faster than any previous CC, CIP, or CID engine ever had before.

THE CID ENGINE PROJECT

We have a vision.

Clean and efficient transportation is widely available. Electricity is plentiful and pollution-free. Space travel is as common as a flight on an airplane is today. Life spreads out from Earth into the Solar System and unto the stars beyond.

As you now know, this is neither a science fiction fantasy nor mere theory. The discoveries made over forty years ago by Bob Cook and continued by his son Rob Cook, Jr., are key to turning this dream into reality.

The future is overdue. We want this technology to benefit everyone. It is time to bring the space age out of our garage and into yours.

To do this, these inventions need to be mass produced to replace the combustion engines we use today. Patent protection of the new concepts (like the CID engine) and refinement will ensure this happens.

By simply reading this book you are helping us to achieve this goal. Spread the word to your family, friends, coworkers, and to the media; encourage them to get a copy, too. Every contribution helps move this project forward faster.

We believe in the power each one of us has to affect the course of the future. The dedication of two men and those closest to them brought this discovery to life. With your assistance, we can bring this astounding technology into *all* our lives.

Act now and look forward.

CID-2 proved a fourth method of inertial propulsion worked.

The first configuration of CID-3 in 2011.

For more information and updates about the CID Engine Project, visit:

http://cforce.cc

Made in United States
North Haven, CT
05 November 2026

21760258R00057